Lecture Notes in Mathematics

Wojciech Banaszczyk

Additive Subgroups of Topological Vector Spaces

Springer-Verlag
Berlin Heidelberg New York
London Paris Tokyo
Hong Kong Barcelona
Budapest

Author

Wojciech Banaszczyk
Institute of Mathematics
Łódź University
Banacha 22
90-238 Łódź, Poland

Mathematics Subject Classification (1980): 11H06, 22-02, 22A10, 22A25, 22B05, 40J05, 43-02, 43A35, 43A40, 43A65, 46A12, 46A25, 46B20, 47B10, 52A43, 60B15

ISBN 3-540-53917-4 Springer-Verlag Berlin Heidelberg New York
ISBN 0-387-53917-4 Springer-Verlag New York Berlin Heidelberg

Printed in Germany

Printing and binding: Druckhaus Beltz, Hemsbach/Bergstr.
2146/3140-543210 - Printed on acid-free paper

PREFACE

In the commutative harmonic analysis there are at least two important theorems that make sense without the assumption of the local compactness of the group and the existence of the Haar measure: the Pontryagin-van Kampen duality theorem and the Bochner theorem on positive-definite functions. The Pontryagin-van Kampen theorem is known to be true e.g. for Banach spaces, products of locally compact groups or additive subgroups and quotients of nuclear Fréchet spaces. The Bochner theorem remains valid for locally convex spaces over p-adic fields, for nuclear locally convex spaces (the Minlos theorem), their subgroups and quotients. These lecture notes are an attempt of clearing up the existing material and of determining the "natural" limits of the applicability of the theory. Pontryagin duality is discussed in chapter 5 and the Bochner theorem in chapter 4.

Our exposition is based on the notion of a nuclear group. Roughly speaking, nuclear groups form the smallest class of abelian topological groups which contains locally compact groups and nuclear spaces and is closed with respect to the operations of taking subgroups, Hausdorff quotients and arbitrary products. The definition and basic properties of nuclear groups are gathered in chapter 3. It turns out that, from the point of view of continuous characters, nuclear groups inherit many properties of locally compact groups.

In chapter 2 we show that the assumption of nuclearity is essential: if a separable Fréchet space E is not nuclear, it contains a discrete additive subgroup K such that the quotient group E/K does not admit any non-trivial continuous unitary representations.

In section 10 we apply nuclear groups to obtain answer to an old problem of S. Ulam on rearrangement of series in topological groups. From the point of view of convergence of series and sequences, nuclear groups inherit many properties of finite dimensional and nuclear spaces.

The characteristic feature of our considerations is their geometrical complexion. The heart of the monograph is section 3 on relations between lattices and n-dimensional ellipsoids in R^n. The main tools used here are the Minkowski convex body theorem and the Korkin-Zolotarev bases. To derive the results of chapters 3-5 from those of section

3, we need only some, rather elementary, topology and topological algebra. The main result of chapter 2 is a consequence of the Minkowski--Hlawka theorem and certain properties of ellipsoids of inertia of convex bodies. The analytic apparatus is made use of to a slight degree only. In that sense, our approach to duality is kept in the spirit of the original geometrical idea of Pontryagin.

This monograph lies on the line of several branches of mathematics, sometimes even quite distant, and the author wishes to thank many persons for their remarks and advice which enabled him to present each particular branch in conformity with the current state of knowledge: S. Kwapień and A. Pełczyński for their help in functional analysis, especially the local theory of Banach spaces; H.W. Lenstra, Jr. for comments on the geometry of numbers; V.M. Kadets for information concerning rearrangement of series, and many others.

Particular thanks are directed to W. Wojtyński for his encouraging suggestions; this work is a development of his ideas.

Łódź, July 1990

CONTENTS

Chapter 1

PRELIMINARIES

In this chapter we establish notation and terminology. We also state some standard facts in a form convenient to us. Section 1 is devoted to abelian topological groups and section 2 to topological vector spaces. In section 3 we give some more or less known facts about additive subgroups of R^n.

1. Topological groups

The groups under consideration will be mostly additive subgroups or quotient groups of vector spaces. Therefore we shall apply the additive notation mainly, denoting the neutral element by 0. Naturally, we shall keep the multiplicative notation for groups of, say, non-zero complex numbers or linear operators. The additive groups of integers and of real and complex numbers will be denoted by Z, R and C, respectively. The multiplicative group of complex numbers with modulus 1 will be denoted by S.

By a <u>character</u> of a group G we mean a homomorphism of G into the group $T := R/Z$. We shall frequently identify T with the interval $(-\frac{1}{2},\frac{1}{2}]$. The canonical projection of R onto T will be denoted by ρ. Thus $\rho(x) = x$ for $x = (-\frac{1}{2},\frac{1}{2}]$. The value of character χ at a point g will be denoted by $\chi(g)$ or, sometimes, by $\langle\chi,g\rangle$.

Now, let G be an abelian topological group (we do not assume topological groups to be Hausdorff). The set of all continuous characters of G, with addition defined pointwise, is an abelian group again. We call it the <u>dual group</u> or the <u>character group</u> of G and denote by $G^\wedge$.

Characters are usually defined as homomorphisms into S. Such a definition is convenient in harmonic analysis, when we consider complex-valued functions "synthesized" of continuous characters (such a situation will take place in chapter 4). However, it leads to the multiplicative notation on $G^\wedge$ which is inconvenient in duality theory when we try to maintain symmetry between G and $G^\wedge$ (especially when we consider topological vector spaces). There are also certain technical reasons for which we have chosen T instead of S.

We shall have to consider various topologies on $G^\wedge$. The dual group endowed with a given topology τ will be denoted $G^{\wedge}_{\tau}$. By $G^{\wedge}_{p}$, $G^{\wedge}_{c}$ and $G^{\wedge}_{pc}$ we shall denote the dual group endowed, respectively, with the topology of uniform convergence on finite, compact and precompact subsets of G (i.e. with the topology of pointwise, compact and precompact convergence). The second one is usually called the compact-open topology.

Now, let A be a subset of G. If χ is a character of G, then we write

$$|\chi(A)| = \sup \{|\chi(g)|: g \in A\}.$$

The set

$$\{\chi \in G^{\wedge} : |\chi(A)| \leq \frac{1}{4}\}$$

is called the <u>polar</u> of A in $G^{\wedge}$; we denote it by A°_{G}. If the meaning of G is clear, we simply write A° instead of A°_{G}. By A°_{p}, A°_{c} and A°_{pc} we denote the set A° endowed with the topology of pointwise, compact and precompact convergence, respectively.

If A is a subgroup of G, then

$$A^{\circ} = \{\chi \in G^{\wedge} : \chi|_{A} \equiv 0\};$$

this follows, for instance, from (1.2). Thus A° is a closed subgroup of $G^{\wedge}_{p}$; we call it the <u>annihilator</u> of A.

A subset A of G is said to be <u>quasi-convex</u> if to each $g \in G \setminus A$ there corresponds some $\chi \in A^{\circ}$ with $|\chi(g)| > \frac{1}{4}$. The set

$$\bigcap_{\chi \in A^{\circ}} \{g \in G : |\chi(g)| \leq \frac{1}{4}\}$$

is evidently the smallest quasi-convex subset of G containing A; we call it the <u>quasi-convex hull</u> of A. We say that G is a <u>locally quasi-convex</u> group if it admits a base at zero consisting of quasi--convex sets. Observe that if G is a Hausdorff locally quasi-convex group, then it admits sufficiently many continuous characters (i.e. continuous characters separate the points of G). Observe also that the polar of any subset of G is a quasi-convex subset of each of the groups $G^{\wedge}_{p}$, $G^{\wedge}_{c}$ and $G^{\wedge}_{pc}$; therefore all the three groups are locally quasi-convex.

(1.1) LEMMA. Let g,h be two elements of an abelian group G. If χ is a character of G such that $|\chi(g)|$, $|\chi(h)|$ and $|\chi(g + h)|$ are less than $\frac{1}{3}$, then $\chi(g + h) = \chi(g) + \chi(h)$.

Proof. One has

$$\chi(g + h) \equiv \chi(g) + \chi(h) \pmod{Z},$$

i.e.

(1) $$\chi(g + h) - \chi(g) - \chi(h) \in Z.$$

From our assumption we obtain

(2) $$|\chi(g + h) - \chi(g) - \chi(h)| \leq |\chi(g + h)| + |\chi(g)| + |\chi(h)| < \frac{1}{3} + \frac{1}{3} + \frac{1}{3} = 1.$$

Now (1) and (2) imply that $\chi(g + h) - \chi(g) - \chi(h) = 0$. ∎

(1.2) LEMMA. Let χ be a character of an abelian group G. Let m be a positive integer and g an element of G, such that $\chi(kg) < \frac{1}{3}$ for $k = 1,\ldots,m$. Then $\chi(mg) = m\chi(g)$.

Proof. By the preceding lemma, for each $k = 1,\ldots,m-1$, we have $\chi((k + 1)g) = \chi(kg) + \chi(g)$. Thus

$$\sum_{k=1}^{m-1} \chi((k + 1)g) = \sum_{k=1}^{m-1} \chi(kg) + \sum_{k=1}^{m-1} \chi(g),$$

which means that $\chi(mg) = m\chi(g)$. ∎

Let A be a subset of an abelian group G. By $\mathrm{gp}\, A$ we denote the subgroup of G generated by A. For each $m = 1,2,\ldots,$ we denote

$$A^m = \{a_1 + \ldots + a_m : a_1,\ldots,a_m \in A\}.$$

(1.3) PROPOSITION. Let G be an abelian topological group. The polars of compact (resp. finite, precompact) subsets of G form a base at zero in $G^\wedge_c$ (resp. in $G^\wedge_p$, $G^\wedge_{pc}$).

Proof. Let U be a neighbourhood of zero in $G^\wedge_c$ (resp. in $G^\wedge_p$, $G^\wedge_{pc}$). There exist an $\varepsilon > 0$ and a compact (resp. finite, precompact) subset Y of G, such that the set $W = \{\chi \in G^\wedge : |\chi(Y)| < \varepsilon\}$ is contained in U. Choose an integer $m > (4\varepsilon)^{-1}$. The set $A = Y^m$ is compact (resp. finite, precompact). By (1.2), for each $\chi \in A^o$, we have $|\chi(Y)| \leq \frac{1}{m}|\chi(A)| \leq \frac{1}{4m} < \varepsilon$. Thus $A^o \subset W$. ∎

By $N_o(G)$ we denote the family of all neighbourhoods of zero in an abelian topological group G (we do not assume neighbourhoods to be open).

(1.4) LEMMA. A character χ of an abelian topological group G is continuous if and only if $\chi \in U^o$ for a certain $U \in \mathcal{N}_o(G)$.

Proof. The necessity of the condition is trivial. To prove the sufficiency, choose any $\varepsilon > 0$. We can find an integer $m > (4\varepsilon)^{-1}$ and then some $W \in \mathcal{N}_o(G)$ with $W^m \subset U$. By (1.2), we have $|\chi(W)| \le \frac{1}{m}|\chi(U)| \le \frac{1}{4m} < \varepsilon$. This means that χ is continuous at zero. ∎

(1.5) PROPOSITION. The polars of neighbourhoods of zero in an abelian topological group G are compact subsets of $G^\wedge_{pc}$.

Proof. Choose any $U \in \mathcal{N}_o(G)$. The group $G^\wedge_p$ is compact because we may identify it with a closed subgroup of the product T^G (see also (1.8)). Since U^o_p is a closed subset of $G^\wedge_p$, it is enough to show that the identity mapping $U^o_p \to U^o_{pc}$ is continuous.

Choose any $\kappa \in U^o$ and let W be a neighbourhood of κ in U^o_{pc}. By (1.3), there is some precompact subset A of G such that

$$W' := (\kappa + A^o) \cap U^o \subset W.$$

Next, we can find some $V \in \mathcal{N}_o(G)$ with $V^3 \subset U$. Since A is precompact, there exist some $g_1,\dots,g_n \in A$ such that

(1) $$A \subset \{g_i\}_{i=1}^n + V.$$

The set

$$W'' = \{\chi \in U^o : |\chi(g_i) - \kappa(g_i)| \le \tfrac{1}{12} \quad \text{for} \quad i = 1,\dots,n\}$$

is a neighbourhood of κ in U^o_p. It remains to show that $W'' \subset W'$.

So, choose any $\chi \in W''$. We have to show that $\chi - \kappa \in A^o$. Take any $g \in A$. In view of (1), we may write $g = g_i + h$ for some $i = 1,\dots,n$ and some $h \in V$. Now, from (1.2) we obtain

$$|\chi(V)| \le \tfrac{1}{3}|\chi(U)| \le \tfrac{1}{12} \quad \text{and} \quad |\kappa(V)| \le \tfrac{1}{3}|\kappa(U)| \le \tfrac{1}{12}.$$

Hence

$$|(\chi - \kappa)(g)| \le |\chi(g_i) - \kappa(g_i)| + |\chi(h)| + |\kappa(h)| \le \tfrac{3}{12} = \tfrac{1}{4}. \quad ∎$$

An abelian group G is called <u>divisible</u> if to each $g \in G$ and each $n = 1,2,\dots$ there corresponds some $h \in G$ with $nh = g$.

(1.6) PROPOSITION. Let H be a subgroup of an abelian group G. Every homomorphism of H into a divisible group can be extended to a homomorphism of G.

For the proof, see e.g. [38], Theorem (A.7).

Let G,H be abelian topological groups. An isomorphism ϕ of G onto H is called a topological isomorphism if ϕ and ϕ^{-1} are continuous. If there is a topological isomorphism of G onto H, then we say that G and H are topologically isomorphic and write $G \simeq H$. An injective homomorphism $\phi : G \to H$ is called a topological embedding if ϕ is a topological isomorphism of G onto the group $\phi(G)$ endowed with the topology induced from H.

(1.7) PROPOSITION. Let $G_1,\dots,G_n$ be abelian topological groups. There is a canonical topological isomorphism between $(G_1 \times \dots \times G_n)^\wedge_c$ and $(G_1)^\wedge_c \times \dots \times (G_n)^\wedge_c$.

This is a standard fact. For the proof, we refer the reader to [38], (23.18) - the assumption there that $G_1,\dots,G_n$ are locally compact is inessential. For infinite products, see (14.11) below.

Let H be a subgroup of an abelian topological group G. We say that H is <u>dually closed</u> in G if to each $g \in G \setminus H$ there corresponds some $\chi \in H^0$ with $\chi(g) \neq 0$ (this is equivalent to the assertion that H is a quasi-convex subset of G). Next, we say that H <u>dually embedded</u> in G if each continuous character of H can be extended to a continuous character of G. Observe that dually closed subgroups are closed. Observe also that each continuous character of H can be extended in a unique way to a continuous character of $\overline{H}$.

Let us recall shortly basic facts concerning the Pontryagin - van Kampen duality theorem. The proofs can be found e.g. in [38], §24. By a compact (resp. locally compact) group we shall mean a group which is compact (resp. locally compact) and separated. Locally compact abelian groups are called LCA groups.

(1.8) PROPOSITION. Let G be an LCA group. Then $G^\wedge_c = G^\wedge_{pc}$ is an LCA group, too, and the evaluation map is a topological isomorphism of G onto $(G^\wedge_c)^\wedge_c$. If H is a closed subgroup of G, then H is dually closed and dually embedded. Moreover, the canonical mappings $G^\wedge_c/H^0_c \to H^\wedge_c$ and $(G/H)^\wedge_c \to H^0_c$ are both topological isomorphism. If G

is compact, $G_c^\wedge$ is discrete. If G is discrete, $G_c^\wedge$ is compact. There are canonical topological isomorphisms $R_c^\wedge \to R$, $Z_c^\wedge \to T$, $T_c^\wedge \to Z$.

(1.9) PROPOSITION. Let G be an LCA group. Then there exist an $n = 0,1,2,\ldots$, a compact group K and a discrete group D, such that G is topologically isomorphic to a closed subgroup of $R^n \times K \times D$.

Proof. Being an LCA group, $G_c^\wedge$ contains an open subgroup $A \cong R^n \times H$ for some $n = 0,1,2,\ldots$ and some compact group H ([69], Theorem 25 or [38], (9.14)). Let $\phi : G^\wedge \to G^\wedge/A$ be the natural projection. Every (abelian) group is a quotient of a free one. So, we can find a free abelian group F and a homomorphism ψ of F onto $G^\wedge/A$. Let $\{f_i\}_{i\in I}$ be a system of free generators of F. For each $i \in I$, choose some $\chi_i \in G^\wedge$ with $\phi(\chi_i) = \psi(f_i)$. Let $\sigma : F \to G^\wedge$ be the homomorphism given by $\sigma(f_i) = \chi_i$ for $i \in I$. We obtain the following commutative diagram:

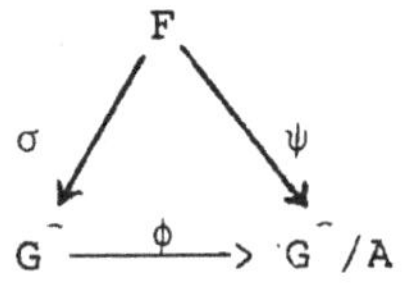

The formula

$$\rho(a,f) = a + \sigma(f) \qquad (a \in A, \quad f \in F)$$

defines a homomorphism $\rho : A \times F \to G^\wedge$. We shall prove that

$$(1) \qquad \rho(A \times F) = G^\wedge.$$

To this end, choose an arbitrary $\chi \in G^\wedge$. Since $\psi(F) = G^\wedge/A$, we can find some $f \in F$ with $\psi(f) = \phi(\chi)$. Then $\phi(\sigma(f)) = \psi(f) = \phi(\chi)$, which means that $a := \chi - \sigma(f) \in \ker\phi = A$. Thus $\chi = a + \sigma(f) = \rho(a,f) \in \rho(A \times F)$, which proves (1).

Let us endow F with the discrete topology. Since $A \times \{0\}$ is an open subgroup of $A \times F$ and ρ is a topological isomorphism (in fact, an identity) of $A \times \{0\}$ onto the open subgroup A of $G_c^\wedge$, it follows that $\rho : A \times F \to G_c^\wedge$ is both continuous and open. Consequently, $G_c^\wedge \cong (A \times F)/\ker\rho$. So, in virtue of (1.8), we have

$$G \cong (G_c^\wedge)_c^\wedge \cong ((A \times F)/\ker\rho)_c^\wedge \cong (\ker\rho)_c^{\circ}.$$

In other words, G is topologically isomorphic to a closed subgroup of

$(A \times F)^{\wedge}_{c}$. From (1.7) we get $(A \times F)^{\wedge}_{c} \simeq A^{\wedge}_{c} \times F^{\wedge}_{c} \simeq R^n \times H^{\wedge}_{c} \times F^{\wedge}_{c}$ and it remains to observe that $F^{\wedge}_{c}$ is compact and $H^{\wedge}_{c}$ discrete. ∎

The completion of an abelian topological group G will be denoted by $\tilde{G}$. We shall identify G with a dense subgroup of $\tilde{G}$. The closures in $\tilde{G}$ of elements of any given base at zero in G form a base at zero in $\tilde{G}$ ([23], Ch. III, §3, Proposition 7).

(1.10) PROPOSITION. Let G_o be a dense subgroup of an abelian topological group G. Let H be the closure in G of a closed subgroup H_o of G_o and let $\phi : G \to G/H$ be the canonical projection. Then the canonical bijection $G_o/H_o \to \phi(G_o)$ is a topological isomorphism of G_o/H_o onto a dense subgroup of G/H.

This is Proposition 21 of [23], Ch. III, §2.

(1.11) PROPOSITION. Let G be an abelian topological group. If G is a k-space, then $G^{\wedge}$ is a complete group.

Proof. The space T^G of continuous mappings from G to T is complete in the compact-open topology ([52], Ch. 7, Theorem 12). It remains to observe that $G^{\wedge}$ is a closed subset of T^G. ∎

(1.12) PROPOSITION. Let G be an abelian group and $\mathcal{B}$ a family of subsets of G satisfying the following conditions:

(a) every member of $\mathcal{B}$ contains zero;

(b) to each $U \in \mathcal{B}$ there corresponds some $V \in \mathcal{B}$ with $-V \subset U$;

(c) to each $U \in \mathcal{B}$ there corresponds some $V \in \mathcal{B}$ with $V+V \subset U$.

Then there exists a unique topology τ on G compatible with the group structure, such that $\mathcal{B}$ is a base at zero for τ.

For the proof, see [23], Ch. III, §1.2.

Let $\{G_i\}_{i\in I}$ be a family of abelian topological groups indexed by a set I. The product of these groups is defined in the usual way; we denote it by $\prod_{i\in I} G_i$. It is evident that the product of a family of locally quasi-convex groups is locally quasi-convex.

Let $\{p_{ij} : G_i \to G_j;\ i,j \in I,\ i \geq j\}$ be an inverse system of topological groups, that is to say, I is a directed set and, for each pair $i,j \in I$ with $i \geq j$, a continuous homomorphism $p_{ij} : G_i \to G_j$ is given, such that $p_{ij} \cdot p_{jk} = p_{ik}$ if $i \geq j \geq k$. We define the limit of this system in the usual way, identifying it with the appropriate subgroup of the product $\prod_{i\in I} G_i$. If I is the set of positive integers, then we speak of an inverse sequence. Naturally, the limit of an inverse system of locally quasi-convex groups is locally quasi--convex. The product $\prod_{i\in I} G_i$ is canonically topologically isomorphic to the limit of the inverse system $\prod_{i\in K} G_i$ where K runs through all finite subsets of I and the projections p_{KL} for $K \supset L$ are defined in the usual way.

The limit G of the inverse system $\{p_{ij}: G_i \to G_j\}$ may be equal to zero. If, however, I is at most countable and all p_{ij} are onto, then also all projections $p_i : G \to G_i$ are onto. Kaplan [50], Lemma 4.6, proved that if $\{p_{ij} : G_i \to G_j\}$ is an inverse sequence of LCA groups such that $p_{ij}(G_i)$ is dense in G_j for all pairs i,j with $i \geq j$, then also $p_i(G)$ is dense in G_i for every i.

Again, let $\{G_i\}_{i\in I}$ be a family of abelian topological groups. Their <u>direct sum</u>, denoted by $\sum_{i\in I} G_i$, is algebraically the subgroup of the product $\prod_{i\in I} G_i$, consisting of finite sequences (that is, an element $(g_i)_{i\in I}$ of $\prod_{i\in I} G_i$ belongs to $\sum_{i\in I} G_i$ if and only if $g_i = 0$ for all but a finite number of indices i). We shall consider on $\sum_{i\in I} G_i$ the asterisk and the rectangular topologies. To describe them, we have to introduce some additional notions.

Let U be a subset of an abelian group G. For each $g \in U$, we define

$$n_U = \sup \{n : kg \in U \quad \text{for} \quad k = 1,\dots,n\}$$

and $g/U = (n_U)^{-1}$. This means, in particular, that $g/U = 0$ if and only if $kg \in U$ for every k.

Let us suppose that, for each $i \in I$, we are given some $U_i \in \mathcal{N}_o(G_i)$. We denote

$$\sum_{i\in I} U_i = \{ (g_i)_{i\in I} \in \sum_{i\in I} G_i : g_i \in U_i \quad \text{for all} \quad i \in I\},$$

$$\sum_{i\in I}{}^{*} U_i = \{(g_i)_{i\in I} \in \sum_{i\in I} U_i : \sum_{i\in I} (g_i/U_i) < 1\}.$$

Let $\mathcal{B}$ be the family of all sets of the form $\sum_{i\in I} U_i$ where $U_i \in \mathcal{N}_o(G_i)$ for every i. Similarly, let $\mathcal{B}^*$ be the family of all sets of the form $\sum_{i\in I}^{*} U_i$ where $U_i \in \mathcal{N}_o(G_i)$ for every i. It follows from (1.12) that there is a unique topology on $\sum_{i\in I} G_i$ compatible with the group structure, for which $\mathcal{B}$ is a base at zero; we call it the <u>rectangular topology</u>. Conditions (a) - (c) of (1.12) are satisfied trivially. Similarly, it follows from (1.12) that there is a unique topology on $\sum_{i\in I} G_i$ compatible with the group structure, for which $\mathcal{B}^*$ is a base at zero; we call it the <u>asterisk topology</u>. The only non-trivial thing here is to verify (1.12) (c) with $\mathcal{B}$ replaced by $\mathcal{B}^*$:

(1.13) LEMMA. Let $\{G_i\}_{i\in I}$ be a family of abelian topological groups and let $\mathcal{B}^*$ be defined as above. Then to each $U \in \mathcal{B}^*$ there corresponds some $V \in \mathcal{B}^*$ with $V + V \subset U$.

To prove (1.13), we need the following simple proposition whose verification is left to the reader:

(1.14) LEMMA. Let U be a zero-containing subset of an abelian group G. Then $g/(U + U) \le \frac{1}{2}(g/U)$ for each $g \in U$. If V is another zero-containing subset of G and $V + V \subset U$, then $(g + h)/U \le \max(g/V, h/V)$ for all $g,h \in V$.

Proof of (1.13). Choose an arbitrary $U \in \mathcal{B}^*$. We have $U = \sum_{i\in I}^{*} U_i$ for some $U_i \in \mathcal{N}_o(G_i)$, $i \in I$. For each $i \in I$, we can find some $V_i \in \mathcal{N}_o(G_i)$ with $V_i^4 \subset U_i$. Set $V = \sum_{i\in I}^{*} V^i$.

Now, take any sequences $(g_i)_{i\in I}$ and $(h_i)_{i\in I}$ belonging to V. From (1.14) we get

$$\sum_{i\in I} [(g_i + h_i)/U_i] \le \frac{1}{2} \sum_{i\in I} [(g_i + h_i)/(V_i + V_i)]$$

$$\le \frac{1}{2} \sum_{i\in I} \max(g_i/V_i, h_i/V_i) \le \frac{1}{2} \sum_{i\in I} [g_i/V_i + h_i/V_i] < 1.$$

Thus $(g_i + h_i)_{i\in I} \in \sum_{i\in I}^{*} U_i$, which means that $V + V \subset U$. ∎

The asterisk topology is, by definition, finer than the rectangular one. For countable direct sums, these two topologies are identical:

(1.15) PROPOSITION. Let $(G_n)_{n=1}^{\infty}$ be a sequence of abelian topological groups. Then the asterisk topology on $\sum_{n=1}^{\infty} G_n$ is equal to the rectangular one.

Proof. Let $U_n \in N_o(G_n)$ for $n = 1,2,\dots$. We have to show that $\sum_{n=1}^{\infty}{}^* U_n$ contains a rectangular neighbourhood of zero. For each $n = 1,2,\dots$, we can find some $V_n \in N_o(G_n)$ with $V_n^2 \subset U_n$. From (1.14) it follows by induction that

$$(g/U_n) \leq 2^{-n}(g/V_n) \quad \text{for all} \quad n = 1,2,\dots \quad \text{and all} \quad g \in V_n.$$

So, if $(g_n)_{n=1}^{\infty} \in \sum_{n=1}^{\infty} V_n$, then

$$\sum_{n=1}^{\infty} (g_n/U_n) \leq \sum_{n=1}^{\infty} 2^{-n}(g_n/V_n) < \sum_{n=1}^{\infty} 2^{-n} = 1.$$

Thus $\sum_{n=1}^{\infty} V_n \subset \sum_{n=1}^{\infty}{}^* U_n$. ∎

In general, the rectangular topology is not equivalent to the asterisk one (consider, for instance, an uncountable direct sum of real lines). In the sequel, speaking of direct sums of topological groups, we shall always assume that they are endowed with the asterisk topology, unless it is explicitly stated otherwise. Notice that if $\{G_i\}_{i\in I}$ is a family of locally convex spaces, then the group $\sum_{i\in I} G_i$ is topologically isomorphic to their locally convex direct sum.

(1.16) PROPOSITION. The direct sum of an arbitrary family of locally quasi-convex groups is locally quasi-convex.

An easy proof is left to the reader.

(1.17) PROPOSITION. Let G be the direct sum of a family $\{G_i\}_{i\in I}$ of Hausdorff abelian groups. For each $i \in I$, let $\pi_i : G \to G_i$ be the canonical projection. If P is a precompact subset of G relative to the asterisk or rectangular topology, then $\pi_i(P) = \{0\}$ for all but finitely many indices i.

Proof. Suppose that P is precompact in the rectangular topology. Set $J = \{i \in I : \pi_i(P) \neq \{0\}\}$. We have to show that J is finite. Suppose the contrary. To each $i \in J$ there corresponds some $g_i \in P$ with $\pi_i(g_i) \neq 0$. Next, there is some $U_i \in N_o(G_i)$ with $\pi_i(g_i) \notin U_i$ because G_i is separated. The set $U = \sum_{i\in I} U_i$ is a rectangular neighbourhood of zero in G. So, there is a finite subset A of P such that $P \subset A + U$ because P is precompact. Since A is finite and consists of finite sequences while J is infinite, it follows that there is an index $j \in J$ such that $\pi_j(A) = \{0\}$. Then

$$\pi_j(P) \subset \pi_j(A + U) = \pi_j(A) + \pi_j(U) = \{0\} + U_j = U_j.$$

On the other hand, we have $g_j \in P$ and $\pi_j(g_j) \notin U_j$, which is a contradiction. ▪

Let $\{p_{ij} : G_i \to G_j;\ i,j \in I,\ i \le j\}$ be a direct system of abelian topological groups, that is to say, I as a directed set and, for each pair $i,j \in I$ with $i \le j$, a continuous homomorphism p_{ij} : $G_i \to G_j$ is defined, such that $p_{ij} \cdot p_{jk} = p_{ik}$ if $i \le j \le k$. Let G be the direct sum of the family $\{G_i\}_{i \in I}$ and let G_o be the subgroup of G generated by all elements of the form

$$g_i - p_{ij}(g_i) \qquad\qquad (i,j \in I;\ i \le j;\ g_i \in G_i)$$

(we treat G_i and G_j as subgroups of G). We define the limit of the system considered as the quotient group G/G_o. When all groups G_i are locally convex spaces, we obtain the usual definition of the inductive limit.

Kaplan [50] defined the limit of the direct system as the group $G/\overline{G_o}$. He proved that if I is countable, the groups G_i are locally compact and all mappings p_{ij} are injective, then G_o is closed ([50], Theorem 8, p. 433).

It is not hard to see that if J is a cofinal subset of I and all groups G_i are locally quasi-convex, then the limit of the system $\{p_{ij} : G_i \to G_j;\ i,j \in I,\ i \le j\}$ may be identified with the limit of the subsystem $\{p_{ij} : G_i \to G_j;\ i,j \in J,\ i \le j\}$. The assumption of local quasi-convexity is essential.

If I is the set of positive integers, then we speak of direct sequences. In view of the last remark, when considering limits of countable direct systems we may restrict ourselves to limits of direct sequences.

The direct sum of a family $\{G_i\}_{i \in I}$ of locally quasi-convex groups is easily seen to be topologically isomorphic to the limit of the direct system $\{p_{KL} : \sum_{i \in K} G_i \to \sum_{i \in L} G_i\}$ where K,L run through finite subsets of I and the embeddings p_{KL} are defined in the usual way.

(1.18) PROPOSITION. Let G be the limit of a direct sequence $\{p_n : G_n \to G_{n+1}\}$ of abelian topological groups, in which all mappings p_n are topological embeddings. Then the topology of G induces original topologies on the groups G_n. Consequently, if all groups G_n are separated, so is G.

Proof. We may assume that $(G_n)_{n=1}^{\infty}$ is an increasing sequence of subgroups of G. Let $\mathcal{B}$ be the family of all sets of the from

$$U_1 + U_2 + \ldots := \bigcup_{n=1}^{\infty} (U_1 + \ldots + U_n)$$

where $U_n \in \mathcal{N}_o(G_n)$ for every n. It follows directly from (1.15) that $\mathcal{B}$ is a base at zero in G.

Fix an index m and choose an arbitrary $V \in \mathcal{N}_o(G_m)$. We are to find some $U \in \mathcal{N}_o(G)$ with $U \cap G_m \subset V$. Naturally, we may assume that $m = 1$. There is some $U_1 \in \mathcal{N}_o(G_1)$ with $U_1 + U_1 \subset V$, and a simple inductive argument allows us to find, for each $n \geq 2$, some $U_n, W_n \in \mathcal{N}_o(G_n)$ with $W_n \cap G_{n-1} \subset U_{n-1}$ and $U_n + U_n \subset W_n$. It remains to show that $G_1 \cap (U_1 + U_2 + \ldots) \subset V$. Set $Y_k = U_1 + \ldots + U_k + U_k$ for $k = 1,2,\ldots$. It is enough to show that $G_1 \cap Y_k \subset V$ for every k. For $k = 1$, this is obvious. For $k > 1$, we use induction:

$$\begin{aligned} G_1 \cap Y_k &\subset G_1 \cap G_{k-1} \cap (U_1 + \ldots + U_{k-1} + W_k) \\ &= G_1 \cap (U_1 + \ldots + U_{k-1} + (W_k \cap G_{k-1})) \\ &\subset G_1 \cap (U_1 + \ldots + U_{k-1} + U_{k-1}) = G_1 \cap Y_{k-1}, \end{aligned}$$

which is contained in V due to the inductive assumption. ▪

A topological vector space is locally convex if and only if it is a Hausdorff locally quasi-convex group (see (2.4)). Komura [55] showed that the limit of an uncountable direct system $\{p_{ij} : E_i \to E_j\}$ of locally convex spaces in which all mappings p_{ij} are topological embeddings need not be locally convex, and even if it is, it need not induce original topologies on the spaces E_i. If, in (1.18), all groups G_n are locally convex spaces, G is a locally convex space, too ([80], Ch. II, (6.4)). If we assume only that all G_n's are Hausdorff locally quasi-convex groups, then probably G need not be locally quasi-convex. See, however, (7.9). Vilenkin [99] considered another topology on the limit of a direct system. Under his definition, the limit of <u>any</u> direct system of abelian topological groups is a locally quasi-convex group.

Let G,H be abelian topological groups and let $\phi : G \to H$ be a continuous homomorphism. Then the formula

$$\langle\psi(\chi),g\rangle = \langle\chi,\phi(g)\rangle \qquad (\chi \in H^{\wedge};\ \ g \in G)$$

defines a homomorphism $\psi : H^{\wedge} \to G^{\wedge}$. We call it the <u>dual</u> <u>homomorphism</u> and denote by $\phi^{\wedge}$. It is clear that $\phi^{\wedge} : H^{\wedge}_{\tau} \to G^{\wedge}_{\tau}$ is continuous when τ is the topology of pointwise (resp. compact, precompact) convergence. If $\{p_{ij} : G_i \to G_j;\ i,j \in I;\ i \geq j\}$ is an inverse system of abelian topological groups, then $\{p^{\wedge}_{ij} : G^{\wedge}_j \to G^{\wedge}_i\}$ is a direct system, and vice versa.

Let $\{G_i\}_{i\in I}$ be a family of abelian topological groups indexed by a set I. Suppose that, for each $i \in I$, a closed subgroup H_i of G_i has been chosen. Let G be the subgroup of the product $\prod_{i\in I} G_i$ consisting of all sequences $(g_i)_{i\in I}$ such that $g_i \in H_i$ for all but finitely many indices i. We topologize G by identifying it with the limit of the inverse system

$$\pi_{KL} : \sum_{i\in K} G_i \times \sum_{i\notin K} (G_i/H_i) \to \sum_{i\in L} G_i \times \sum_{i\notin L} (G_i/H_i)$$

where K,L are finite subsets of I with $K \supset L$, and π_{KL} is the canonical projection. Endowed with this topology, G will be called the <u>reduced</u> <u>product</u> of groups G_i relative to subgroups H_i and denoted by $\sum_{i\in I} (G_i : H_i)$. It is clear that the topology of G induces original topologies on the subgroups G_i. Notice that if $H_i = G_i$ for almost all i, then $G = \prod_{i\in I} G_i$; if $H_i = \{0\}$ for almost all i, then $G = \sum_{i\in I} G_i$. More precisely, if H is the subgroup of G consisting of all $(g_i)_{i\in I}$ such that $g_i \in H_i$ for every i, then H has the usual product topology and G/H is topologically isomorphic to the direct sum $\sum_{i\in I} (G_i/H_i)$. Since inverse limits and direct sums of locally quasi-convex groups are locally quasi-convex, G is locally quasi-convex provided that so are all groups G_i and G_i/H_i.

(1.19) PROPOSITION. For each $i \in I$, let $\psi_i : G \to G_i$ be the canonical projection. A subset X of G is precompact if and only if $\psi_i(X)$ is precompact in G_i for all i, and $\psi_i(X) \subset H_i$ for almost all i.

This is a direct consequence of (1.17) and the definition of G.

(1.20) PROPOSITION. If I is at most countable and G locally quasi-convex, then G may be identified with the limit of the direct system

$$\phi_{KL} : \prod_{i\in K} G_i \times \prod_{i\notin K} H_i \to \prod_{i\in L} G_i \times \prod_{i\notin L} H_i$$

where K,L are finite subsets of I with $K \subset L$ and ϕ_{KL} is the canonical embedding.

The proof is left to the reader as an exercise.

Let G be a topological group (abelian or not). By a <u>representation</u> of G in a real or complex Banach space E we mean a homomorphism of G into the group $GL(E)$ of automorphisms of E. We say that E is the space of the representation. The operator being the value of a representation Φ at a point $g \in G$ will be denoted by $\Phi(g)$ or Φ_g. We say that Φ is <u>faithful</u> if $\Phi_g \neq \mathrm{id}$ for $g \neq 1$. A representation $\Phi : G \to GL(E)$ is called <u>weakly</u> (resp. <u>strongly</u>) <u>continuous</u> if, for each fixed $u \in E$, the mapping $g \to \Phi_g u$ is continuous in the weak (resp. strong) topology on E. We say that Φ is <u>uniformly continuous</u> if the mapping $g \to \Phi_g$ is continuous in the norm topology on $GL(E)$.

We say that Φ is a <u>cyclic representation</u> if there exists a vector $u \in E$ such that the linear subspace spanned over the vectors $\Phi_g u$, $g \in G$, is dense in E; then u is called a <u>cyclic vector</u> of Φ. A subspace M of E is called <u>invariant</u> (relative to Φ) if $\Phi_g(M) \subset M$ for each $g \in G$. If the only invariant subspaces of E are E itself and $\{0\}$, we say that Φ is an <u>irreducible representation</u>.

Let H be a complex Hilbert space. A representation $\Phi : G \to GL(H)$ is called <u>unitary</u> if all operators Φ_g, $g \in G$, are unitary. A unitary representation is weakly continuous if and only if it is strongly continuous; then we call it simply a <u>continuous unitary representation</u>.

if Φ is an irreducible unitary representation of an abelian group in a complex Hilbert space H, then $\dim H = 1$. Since the multiplicative group of non-zero complex numbers is topologically isomorphic to $R \times T$, it follows that the study of irreducible unitary representations of abelian groups can be reduced to the study of their characters.

Let Φ_1,Φ_2 be two representations of a group G in Banach spaces E_1,E_2, respectively. We say that these representations are equivalent if there is a topological isomorphism $\Theta : E_1 \to E_2$ such that $\Phi_1\Theta = \Theta\Phi_2$. If E_1,E_2 are complex Hilbert spaces and the representations Φ_1 and Φ_2 are unitary, then we say that Φ_1 is unitarily equivalent to Φ_2 if there is a unitary isomorphism $\Theta : E_1 \to E_2$ with $\Phi_1\Theta = \Theta\Phi_2$.

Let $\{H_i\}_{i\in I}$ be a family of complex Hilbert spaces. Their Hilbert sum, denoted by $\bigoplus_{i\in I} H_i$, is the subspace of the product $\prod_{i\in I} H_i$, consisting of all sequences $(u_i)_{i\in I}$ with $\sum_{i\in I} \|u_i\|^2 < \infty$. The inner product in $\bigoplus_{i\in I} H_i$ is given by the formula

$$((u_i),(v_i)) = \sum_{i\in I} (u_i,v_i).$$

Let G be a topological group and suppose that, for each $i \in I$, a continuous unitary representation Φ_i of G in H_i is given. Then the formula

$$\Phi(g)\cdot(u_i)_{i\in I} = (\Phi_i(g)\cdot u_i)_{i\in I}$$

defines a continuous unitary representation Φ of G in the space $\bigoplus_{i\in I} H_i$; we call Φ the Hilbert sum of the representations Φ_i and denote it by $\bigoplus_{i\in I} \Phi_i$.

(1.21) PROPOSITION. Let Φ be a unitary representation of a group G in a complex Hilbert space H. Then there exists a family $\{H_i\}_{i\in I}$ of closed, invariant, pairwise orthogonal subspaces of H such that $\bigcup_{i\in I} H_i$ is linearly dense in H and, for each $i \in I$, the representation Φ_i of G in H_i given by

$$\Phi_i(g)\cdot u_i = \Phi(g)\cdot u_i \qquad (g \in G;\ \ u_i \in H_i)$$

is cyclic. In other words, every unitary representation is unitarily equivalent to the Hilbert sum of cyclic representations.

This is Theorem (21.14) of [38].

A complex-valued function ϕ on a group G is said to be positive-definite (shortly, p.d.) if, for each $n = 1,2,\ldots,$ it satisfies the condition

$$\sum_{i=1}^{n} \sum_{j=1}^{n} \lambda_i \bar{\lambda}_j \phi(g_i^{-1} g_j) \geq 0$$

for all $\lambda_1, \ldots, \lambda_n \in C$ and $g_1, \ldots, g_n \in G$. Every linear combination of characters, with non-negative coefficients, is a p.d. function on G (see [38], (32.9)).

(1.22) PROPOSITION. Let ϕ be a p.d. function on a group G. Then

(a) $\phi(0) \geq 0$;

(b) $|\phi(g)| \leq \phi(0)$ for all $g \in G$;

(c) $|\phi(g) - \phi(h)|^2 \leq 2\phi(0)[\phi(0) - \text{Re } \phi(g-h)]$ for all $g, h \in G$.

This is a direct consequence of the definition of a p.d. function (cf. [38], (32.4)).

(1.23) PROPOSITION. Let Φ be a unitary representation of a group G in a Hilbert space H. For each $u \in H$, the function $\phi(g) = (\Phi_g u, u)$ is positive-definite. If G is a topological group and Φ is continuous, then so is ϕ. If ϕ is continuous and u is a cyclic vector of Φ, then Φ is continuous.

For the proof, see [38], (32.8) (a), (b) and (f).

(1.24) PROPOSITION. Let ϕ be a continuous p.d. function on a topological group G. Then there exists a continuous cyclic unitary representation Φ of G with a cyclic vector u such that $(\Phi_g u, u) = \phi(g)$ for all $g \in G$.

This follows from [38], (32.3) and (32.8) (f).

(1.25) PROPOSITION. Let Φ, Ψ be two continuous unitary representations of a topological group G with cyclic vectors u, w, respectively. If $(\Phi_g u, u) = (\Psi_g w, w)$ for all $g \in G$, then Φ and Ψ are unitarily equivalent.

For the proof, see [18], Proposition 3, p. 146 or [38], (32.8) (b).

(1.26) NOTE. The material of this section is standard.

2. Vector spaces

All vector spaces occurring are assumed to be real. The only exceptions are complex spaces connected with representations of groups and vector spaces over ultrametric fields, considered in section 7. For the terminology (but not for the notation) concerning topological vector spaces we refer the reader to Schaefer's book [80]. Locally convex spaces (in particular, all nuclear spaces) are meant to be separated.

Let E be a vector space. By $E^{\#}$ we denote the space of all linear functionals on E. If E is a topological vector space, the space of all continuous linear functionals on E is denoted by E^*; endowed with the topology of uniform convergence on bounded, compact and precompact sets, it is denoted by E_b^*, E_c^* and E_{pc}^*, respectively. The dimension of E is denoted by $\dim E$. If A is a subset of E, then $\operatorname{span} A$ denotes the linear subspace of E spanned over A and $\operatorname{conv} A$ is the convex hull of A.

(2.1) NOTE. It is convenient to set $\operatorname{span} \emptyset = \{0\}$. Consequently, if $u_1, u_2, \ldots$ is a sequence in E, the symbol $\operatorname{span} \{u_i\}_{i<k}$ for $k = 1$ should be meant as $\{0\}$.

Every vector space may be treated as an additive group. Similarly every topological vector space may be treated as a topological group.

(2.2) LEMMA. Let U be a symmetric, radial and absorbing subset of a vector space E and let χ be a character of E such that $|\chi(U)| < \frac{1}{3}$. Then there exists exactly one $f \in E^{\#}$ with $\rho f = \chi$. For each $u \in U$, one has $f(u) = \chi(u)$; consequently,

$$\sup \{|f(u)| : u \in U\} = |\chi(U)| < \tfrac{1}{3}.$$

Proof. The uniqueness of f is obvious. We shall prove the existence. Given $u \in E$, we can find a positive integer n with $\frac{u}{n} \in U$. Set $f(u) = n\chi(\frac{u}{n})$. This definition does not depend on the choice of n. Indeed, suppose that m is some other positive integer with $\frac{u}{m} \in U$. By (1.2), we have $\chi(m \frac{u}{mn}) = m\chi(\frac{u}{mn})$ and $\chi(n \frac{u}{mn}) = n\chi(\frac{u}{mn})$. Hence

$$m\chi(\tfrac{u}{m}) = mn\chi(\tfrac{u}{mn}) = n\chi(\tfrac{u}{n}).$$

It remains to show that f is linear. Take any $u, w \in E$. There is some m such that $\frac{u}{m}, \frac{w}{m}$ and $\frac{u+w}{m}$ all belong to U. Then, by (1.1),

$$\chi(\frac{u+w}{m}) = \chi(\frac{u}{m}) + \chi(\frac{w}{m}),$$

i.e. $\frac{1}{m}f(u + w) = \frac{1}{m}f(u) + \frac{1}{m}(w)$, which proves that f is an additive mapping. Now, take any scalar $\lambda \in R$. Let $[x]$ and $\{x\}$ denote the integer and the fractional parts of x, respectively. Since f is additive, for each positive integer p, we have

$$f(\lambda u) = f(\frac{1}{p}p\lambda u) = \frac{1}{p}f(p\lambda u)$$
$$= \frac{1}{p}f([p\lambda]u + \{p\lambda\}u) = \frac{1}{p}[p\lambda]f(u) + \frac{1}{p}f(\{p\lambda\}u).$$

The latter expression tends to $\lambda f(u)$ as $p \to \infty$. Thus $f(\lambda u) = f(u)$. ∎

If f is a continuous linear functional on a topological vector space E, then ρf is evidently a continuous character of E. The mapping $f \to \rho f$ is a homomorphism of E^* into $E^\wedge$; we shall denote it by ρ_E.

(2.3) PROPOSITION. Let E be a topological vector space. Then ρ_E is an algebraic isomorphism of E^* onto $E^\wedge$; it is a topological isomorphism of E_c^* onto $E_c^\wedge$.

Proof. That ρ_E maps E^* onto $E^\wedge$ follows directly from (2.2). The continuity is trivial. To prove that ρ_E is an open mapping of E_c^* onto $E_c^\wedge$, take any $U \in N_o(E_c^*)$. There is a compact subset K of E such that

$$K^0 := \{f \in E^* : |f(u)| \leq 1 \text{ for all } u \in K\} \subset U.$$

The set

$$H = \{tu \in E : t \in [-1,1] \text{ and } u \in K\}$$

is compact, so that $H^o \in N_o(E_c^\wedge)$. It remains to prove that $\rho_E(K^0) \supset H^o$.

So, take any $\chi \in H^o$. There is some $f \in E^*$ with $\rho_E(f) = \chi$. Due to (2.2), we have

$$\sup\{|f(u)| : u \in H\} = |\chi(H)| \leq \frac{1}{4},$$

which means that $f \in K^0$. ∎

(2.4) PROPOSITION. A topological vector space E is locally convex if and only if it is a Hausdorff locally quasi-convex group.

Proof. Let E be a locally convex space. Take any symmetric, closed, convex $U \in N_o(E)$. Next, take any $v \in E \setminus U$. By the Hahn--Banach theorem, there is some $f \in E^*$ with $\sup\{f(u) : u \in U\} < f(v)$. Therefore we can find a positive number c such that

$$\sup\{f(u) : u \in U\} < c < f(v) < 3c.$$

Set $h = \frac{f}{4c}$. Since U is symmetric, we have

$$\sup\{|h(u)| : u \in U\} < \tfrac{1}{4} < h(v) < \tfrac{3}{4}.$$

Then $\rho h \in E^{\wedge}$ and

$$\sup\{|\rho h(u)| : u \in U\} < \tfrac{1}{4} < |\rho h(v)|,$$

which means that v does not belong to the quasi-convex hull of U. Since $v \in E \setminus U$ was arbitrary, it follows that U is quasi-convex. This proves that E is a locally quasi-convex group because symmetric, closed, convex neighbourhoods of zero form a base at zero in E.

Now, suppose that a topological vector space E is a Hausdorff locally quasi-convex group. For each subset A of E^*, denote

$$A^o = \{u \in E : f(u) \in [-\tfrac{1}{4},\tfrac{1}{4}] + Z \text{ for all } f \in A\},$$

$$A_o = u \in E : f(u) \in [-\tfrac{1}{4},\tfrac{1}{4}] \text{ for all } f \in A\}.$$

It follows from (2.3) that every quasi-convex subset of E has the form A^o for some $A \subset E^*$. So, there exists a family $\mathcal{A}$ of subsets of E^* such that that the family $\{A^o : A \in \mathcal{A}\}$ constitutes a base at zero in E. To prove that E is a locally convex space, it suffices to show that $\{A_o : A \in \mathcal{A}\}$ is such a base, too.

Choose any $A \in \mathcal{A}$. We have to show that $A_o \in N_o(E)$. Since E is a topological vector space, there is some radial $U \in N_o(E)$ contained in A^o. Then $U \subset A_o$. ∎

(2.5) PROPOSITION. Let K be a subgroup of a topological vector space E. Then the set $\bigcap\{\chi^{-1}(0) : \chi \in K^o\}$ is equal to the weak closure of K. Hence K is dually closed in E if and only if it is weakly closed; it is weakly dense if and only if $(E/K)^{\wedge} = \{0\}$.

Proof. Let $\overline{K}^w$ denote the weak closure of K. Take any $u \in \overline{K}^w$ and $\chi \in K^o$. By (2.3), we have $\chi = \rho f$ for some $f \in E^*$. Then $f(K) \subset \ker \rho = Z$. This implies that $f(u) \in Z$ because $u \in \overline{K}^w$. Hence $\chi(u) = \phi f(u) = 0$ and, consequently, $\overline{K}^w \subset \bigcap\{\chi^{-1}(0) : \chi \in K^o\}$.

To prove the opposite inclusion, choose any $u \in E \setminus \overline{K}^w$. There is a weak neighbourhood of u in E disjoint from K. In other words, there are some $f_1,\dots,f_n \in E^*$ and $\varepsilon > 0$ such that if $|f_i(v-u)| < \varepsilon$ for $i = 1,\dots,n$, then $v \notin K$. Consider the continuous linear operator $F = (f_1,\dots,f_n) : E \to R^n$. The set

$$\{(x_1,\dots,x_n) \in R^n : |x_i - f_i(u)| < \varepsilon \quad \text{for} \quad i = 1,\dots,n\}$$

is a neighbourhood of $F(u)$ in R^n disjoint from $F(K)$. Thus $F(u) \notin \overline{F(K)}$. According to (1.8), we can find some $\kappa \in (R^n)^\wedge$ with $\kappa(\overline{F(K)}) = \{0\}$ and $\kappa(F(u)) \neq 0$. Then $\kappa F \in K^o$ and $(\kappa F)(u) \neq 0$. ■

A subset X of a vector space is called <u>radial</u> if $tu \in X$ for all $u \in X$ and $t \in (0,1)$. Let X,Y be two radial subsets of a vector space E. We write $X \prec Y$ if X is absorbed by Y, i.e. if $X \subset tY$ for some $t > 0$. Suppose that $X \prec Y$. For each linear subspace L of E, we denote

$$d(X,Y;L) = \inf \{t > 0 : X \subset tY + L\}.$$

Next, for each $k = 1,2,\dots,$ we denote

$$d_k(X,Y;E) = \inf_L d(X,Y;L)$$

where the infimum is taken over all linear subspaces L of E with $\dim L < k$. If E is contained in some other vector space F, then $d_k(X,Y;F) = d_k(X,Y;E)$. Therefore we may simply write $d_k(X,Y)$ instead of $d_k(X,Y;E)$. The numbers $d_k(X,Y)$ are called <u>Kolmogorov diameters</u> of X with respect to Y. We take the following natural convention: for $t > 0$, the expression "$d_k(X,Y) < t$" should be read as "$X \prec Y$ and $d_k(X,Y) < t$".

(2.6) LEMMA. Let X,Y,Z be three radial subsets of some vector space.

(a) If Z is convex and $X,Y \prec Z$, then

$$d_{k+l-1}(X + Y,Z) \le d_k(X,Z) + d_l(Y,Z) \qquad (k,l = 1,2,\dots).$$

(b) If $X \prec Y \prec Z$, then

$$d_{k+l-1}(X,Z) \le d_k(X,Y)d_l(Y,Z).$$

For (b), see the proof of Lemma 7.1.2 in [79]. The proof of (a) is similar and we leave it to the reader.

(2.7) LEMMA. Let $(E_n)_{n=1}^{\infty}$ be a sequence of vector spaces and let m be a positive integer. Suppose that, for each $n = 1,2,\ldots,$ we are given radial subsets X_n, Y_n of E_n such that

(1) $$d_k(X_n,Y_n) < 2^{-mn}k^{-m} \qquad (k = 1,2,\ldots).$$

Then $X := \prod_{n=1}^{\infty} X_n$ and $Y := \prod_{n=1}^{\infty} Y_n$ are radial subsets of the space $E := \prod_{n=1}^{\infty} E_n$, and $d_k(X,Y) \le k^{-m}$ for every k. If E_o is the subspace of E consisting of finite sequences (that is, $E_o = \sum_{n=1}^{\infty} E_n$), then $d_k(X \cap E_o, Y \cap E_o) \le k^{-m}$ for every k.

Proof. Let $[x]$ denote the integer part of x. Fix an arbitrary $k = 1,2,\ldots$. If $l = [k2^{-n}]$, then, by (1),

$$d_{l+1}(X_n,Y_n) \le 2^{-mn}(l+1)^{-m} < 2^{-mn}(k2^{-n})^{-m} = k^{-m}.$$

So, for each $n = 1,2,\ldots,$ we can find a subspace L_n of E_n with $\dim L_n \le [k2^{-n}]$ and $X_n \subset k^{-m}Y_n + L_n$. Then $L := \sum_{n=1}^{\infty} L_n$ is a subspace of E_o and we have

$$\dim L = \sum_{n=1}^{\infty} \dim L_n \le \sum_{n=1}^{\infty} [k2^{-n}] < \sum_{n=1}^{\infty} k2^{-n} = k.$$

It remains to observe that $X \subset k^{-m}Y + L$ and $X \cap E_o \subset k^{-m}(Y \cap E_o) + L$. ∎

(2.8) LEMMA. Let E,F be some vector spaces and let X,Y be radial subsets of E with $X \prec Y$.

(a) For each linear operator $\Phi : E \to F$, one has

$$d_k(\Phi(X),\Phi(Y)) \le d_k(X,Y) \qquad (k = 1,2,\ldots).$$

(b) For each linear operator $\Psi : F \to E$ with $\Psi(F) = E$, one has

$$d_k(\Psi^{-1}(X),\Psi^{-1}(Y)) = d_k(X,Y) \quad (k = 1,2,\ldots).$$

This is a direct consequence of the definition of Kolmogorov diameters.

Let E be a normed space. The distance of a point $u \in E$ to a subset A of E is denoted by $d(u,A)$. The closed unit ball in E is denoted by B_E or, sometimes, by $B(E)$. We say that E is a <u>unitary space</u> if its norm is defined by an inner product. The inner product of vectors $u,w \in E$ is denoted by (u,w).

Let $\Phi : E \to F$ be a bounded linear operator acting between normed spaces. For each $k = 1,2,\ldots,$ we write

$$d_k(\Phi : E \to F) = d_k(\Phi(B_E),B_F).$$

The numbers $d_k(\Phi : E \to F)$ are called the <u>Kolmogorov numbers</u> of Φ. In general, they depend not only on Φ, but also on F. For example, $d_k(id : l^1 \to c_o) = 1$, while $d_k(id : l^1 \to l^\infty) = \frac{1}{2}$ for $k = 2,3,\ldots$ (see [76], 11.11.9 and 11.11.10). If the meaning of F is clear from the context, we simply write $d_k(\Phi)$ instead of $d_k(\Phi : E \to F)$ (see also (2.10)). The following lemma is an immediate consequence of definitions.

(2.9) LEMMA. Let $\chi : E' \to E$, $\Phi : E \to F$ and $\Psi : F \to F'$ be bounded operators acting between normed spaces. Then

(a) $\|\Phi\| = d_1(\Phi : E \to F) \geq d_2(\Phi : E \to F) \geq \ldots \geq 0;$

(b) $d_k(\Psi\Phi\chi : E' \to F') \leq \|\Psi\|\|\chi\| d_k(\Phi : E \to F) \quad (k = 1,2,\ldots).$

(2.10) LEMMA. Let $\Phi : E \to F$ be a bounded linear operator. If F is a subspace of some unitary space F', then

$$d_k(\Phi : E \to F) = d_k(\Phi : E \to F') \qquad (k = 1,2,\ldots).$$

An easy proof is left to the reader.

Let X be a convex, absorbing subset of a vector space E. The mapping $u \to \inf\{t > 0 : u \in tX\}$ is called the <u>Minkowski functional</u> of X; it is a seminorm if and only if X is symmetric.

Now, let p be a seminorm on E. We denote $B_p = \{u \in E : p(u) \leq 1\}$. Sometimes we shall write $B(p)$ instead of B_p. The quotient space $E/p^{-1}(0)$ will be denoted by E_p and the natural projection $E \to E_p$ by ψ_p. We shall always consider on E_p the canonical norm given by

$\| \psi_p(u) \| = p(u)$ for $u \in E$. Thus $\psi_p(B_p) = B(E_p)$. We say that p is a <u>pre-Hilbert</u> <u>seminorm</u> if

$$p^2(u + w) + p^2(u - w) = 2p^2(u) + 2p^2(w)$$

for all $u,w \in E$. This holds if and only if the norm on E_p satisfies the parallelogram identity, i.e. if and only if E_p is a unitary space. The following lemma is a direct consequence of our definitions.

(2.11) **LEMMA.** Let $\Phi : E \to F$ be a linear operator between vector spaces. If p is a pre-Hilbert seminorm on E, then $\Phi(B_p)$ is an absorbing subset of the space $\Phi(E)$ and the Minkowski functional of $\Phi(B_p)$ is a pre-Hilbert seminorm on $\Phi(E)$.

Let us suppose that p,q are two seminorms on a vector space E, such that $B_p \prec B_q$, i.e. such that $q \le cp$ for a certain $c > 0$. The canonical operator from E_p to E_q will be denoted by Λ_{pq}. We have have the following commutative diagram:

$$\begin{array}{ccc} E & \xrightarrow{\ \ id\ \ } & E \\ \downarrow \psi_p & & \downarrow \psi_q \\ E_p & \xrightarrow{\ \ \Lambda_{pq}\ \ } & E_q . \end{array}$$

(2.12) **LEMMA.** Let p,q be two seminorms on a vector space E, with $B_p \prec B_q$. Then $d_k(B_p,B_q) = d_k(\Lambda_{pq} : E_p \to E_q)$ for each $k = 1,2,\ldots$.

The proof is quite easy and we leave it to the reader.

(2.13) **LEMMA.** Let p be a seminorm on a vector space E and M a linear subspace of E. If q is a pre-Hilbert seminorm on E and $B_p \prec B_q$, then

$$d_k(M \cap B_p, M \cap B_q) \le d_k(B_p,B_q) \qquad (k = 1,2,\ldots).$$

Proof. Let r,s be the restrictions to M of p,q, respectivelyly. We may identify M_r and M_s with a subspace of E_p and E_q, respectively. Accordingly, we may treat $\Lambda_{rs} : M_r \to M_s$ as a restriction of $\Lambda_{pq} : E_p \to E_q$. Applying (2.12), (2.10) and (2.9) (b), for each $k = 1,2,\ldots,$ we have

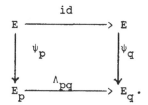

$$d_k(M \cap B_p, M \cap B_q) = d_k(B_r,B_s) = d_k(\Lambda_{rs} : M_r \to M_s)$$
$$= d_k(\Lambda_{rs} : M_r \to E_q) \le d_k(\Lambda_{pq} : E_p \to E_q)$$
$$= d_k(B_p,B_q). \quad \blacksquare$$

The assumption that q is a pre-Hilbert seminorm is essential. For example, $d_k(B(l^1),B(l^\infty)) = \frac{1}{2}$ for $k = 2,3,\ldots,$ while

$$d_k(c_o \cap B(l^1),c_o \cap B(l^\infty)) = d_k(B(l^1),B(c_o)) = 1$$

for $k = 1,2,\ldots,$ (see [76], 11.11.9 and 11.11.10).

(2.14) LEMMA. To each $m = 2,3,\ldots$ there corresponds a constant $c_m > 0$ with the following property: if X,Y are two symmetric, convex subsets of some vector space, with $d_k(X,Y) \le k^{-m}$ for every k, then there are pre-Hilbert seminorms p,q on span X with $X \subset B_p$ $B_q \subset Y$ and $d_k(B_p,B_q) \le c_m \cdot k^{-m+2}$ for every k.

The proof can be obtained by standard methods used in the theory of bounded operators in Banach spaces. It is rather long, but presents no serious difficulties. One may apply, for instance, theorem 8.4.2 of [75].

(2.15) LEMMA. Let p,q be two pre-Hilbert seminorms on a vector space E, with $d_k(B_p,B_q) \to 0$. Let $(a_k)_{k=1}^\infty$ and $(b_k)_{k=1}^\infty$ be two non-increasing sequences of positive numbers, such that $d_k(B_p,B_q) \le a_k \cdot b_k$ for every k. Then there exists a pre-Hilbert seminorm r on E such that $d_k(B_p,B_r) \le a_k$ and $d_k(B_r,B_q) \le b_k$ for every k.

This is an easy consequence of the spectral theorem for compact operators. The details of the proof are left to the reader.

(2.16) LEMMA. Let p,q be two pre-Hilbert seminorms on a vector space E, with $B_p \prec B_q$. Let us denote

$$B_p^0 = \{f \in E^\# : |f(u)| \le 1 \text{ for all } u \in B_p\},$$

$$B_q^0 = \{f \in E^\# : |f(u)| \le 1 \text{ for all } u \in B_q\}.$$

Then $d_k(B_q^0,B_p^0) = d_k(B_p,B_q)$ for every k.

Proof. Consider the linear mapping $\Gamma : E_p \to E$ given by

$$\Gamma(f)(u) = f(\psi_p(u)) \qquad (f \in E_p;\ \ u \in E).$$

It is not hard to verify that $\Gamma^{-1}(B_p^0) = B(E_p^*)$ and $\Gamma^{-1}(B_q^0) = \Lambda_{pq}^*(B(E_p^*))$. Naturally, we may treat Γ as a mapping onto $\Gamma(E_p^*) = \operatorname{span} B_p^0$. Then, by (2.8) (b),

$$\begin{aligned} d_k(B_q^0, B_p^0) &= d_k(\Gamma^{-1}(B_q^0), \Gamma^{-1}(B_p^0)) \\ &= d_k(\Lambda_{pq}^*(B(E_q^*), B(E_p^*))) = d_k(\Lambda_{pq}^* : E_q^* \to E_p^*) \end{aligned}$$

for every k. It is well known that $d_k(\Lambda_{pq}^* : E_q^* \to E_p^*) = d_k(\Lambda_{pq} : E_p \to E_q)$ (see e.g. [76], 11.7.8). Hence, by (2.12), for each $k = 1,2,\ldots,$ we have

$$d_k(B_q^0, B_p^0) = d_k(\Lambda_{pq} : E_p \to E_q) = d_k(B_p, B_q). \quad ■$$

A locally convex space E is said to be <u>nuclear</u> if to each convex $U \in N_o(E)$ there corresponds some convex $W \in N_o(E)$ such that $d_k(W,U) \leq \frac{1}{k}$ for every k.

(2.17) PROPOSITION. Let E be a nuclear space. Choose any $c > 0$ and $m = 1,2,\ldots$. Then to each convex $U \in N_o(E)$ there corresponds some symmetric and convex $W \in N_o(E)$ such that $d_k(W,U) \leq ck^{-m}$ for every k.

This is an easy consequence of (2.6) (b) (cf. the proof of Proposition 7.1.1. in [79]).

(2.18) NOTE. The material of this section is standard.

3. Geometry of numbers

By R^n we denote the n-dimensional euclidean space with the usual norm. The closed unit ball in R^n is denoted by B_n; there will be no possibility of confusion with the symbol B_p where p is a seminorm. The n-dimensional Lebesgue measure on R^n is denoted by vol_n, and the measure of B_n by ω_n. We have

$$\omega_n = \mathrm{vol}_n\,(B_n) = \pi^{n/2}[\Gamma(1 + n/2)]^{-1} \sim (\tfrac{2\pi e}{n})^{n/2}.$$

Throughout the section, D is an n-dimensional ellipsoid in R^n with centre at zero and principal semiaxes $\xi_1 \leq \ldots \leq \xi_n$. Thus

$d_k(B_n, D) = \xi_k^{-1}$ for $k = 1, \dots, n$.

(3.1) PROPOSITION. Let K be a closed subgroup of R^n. Denote by K_o the maximal linear subspace contained in K. If $K \neq K_o$, then there exist linearly independent vectors $a_1, \dots, a_m$, all orthogonal to K_o, such that

$$K = K_o + \{k_1 a_1 + \dots k_m a_m : k_1, \dots, k_m \in Z\}.$$

The proof can be found, for instance, in [23], Ch. VII, §1, n° 2.

Let K be a closed subgroup of R^n and K_o the maximal linear subspace contained in K. It follows from (3.1) that every component of K has the form $u + K_o$ for a certain $u \in K$. The subspace K_o will be called the zero component of K.

By a lattice in R^n we mean an additive subgroup generated by n linearly independent vectors. According to (3.1), lattices may be defined as n-dimensional discrete subgroups of R^n. Let L be a lattice generated by vectors $u_1, \dots, u_n$; we say that the system $(u_k)_{k=1}^n$ is a basis of L. Let $(e_k)_{k=1}^n$ be an orthonormal basis in R^n and let $\Psi : R^n \to R^n$ be the linear operator given by $\Psi e_k = u_k$. The quantity $|\det \Psi|$ is called the determinant of L and denoted by $d(L)$; it does not depend on the choice of a basis. The set

$$\{u \in R^n : (u,w) \in Z \text{ for all } w \in L\}$$

is a lattice, too. We call it the polar lattice and denote by L^*. One has $d(L^*) = [d(L)]^{-1}$ and $L^{**} = L$. All these facts are standard. The proofs can be found, for example, in [28] or [33].

(3.2) LEMMA. Let L be a lattice in R^n with $L \cap D = \{0\}$. Then there exists some $u \in L^*$ with

(1) $$0 < \|u\| \le n(\xi_1 \dots \xi_n)^{-1/n}.$$

Proof. Suppose the contrary. Then $L^* \cap (rB_n) = \{0\}$ for a certain $r > n(\xi_1 \dots \xi_n)^{-1/n}$. So, by virtue of the fundamental Minkowski theorem (see e.g. [33], Theorem 1 on p. 123), we have

$$d(L^*) \ge 2^{-n} \operatorname{vol}_n (rB_n) = 2^{-n} r^n \omega_n \ge r^n n^{-n/2}.$$

On the other hand, from the Minkowski theorem and the assumption that $L \cap D = \{0\}$ we get

$$d(L) \geq 2^{-n} \operatorname{vol}_n (D) = 2^{-n} \omega_n \xi_1 \dots \xi_n \geq n^{-1/2} \xi_1 \dots \xi_n .$$

Hence $d(L)\, d(L^*) \geq r^n n^{-n} \xi_1 \dots \xi_n$. Since $d(L) d(L^*) = 1$, it follows that $r \leq n(\xi_1 \dots \xi_n)^{-1/n}$, which is a contradiction. ▪

(3.3) LEMMA. Let M be an $(n - 1)$-dimensional subspace of R^n, $n \geq 2$. If $\eta_1 \leq \dots \leq \eta_{n-1}$ are the principal semiaxes of $D \cap M$, then $\xi_k \leq \eta_k \leq \xi_{k+1}$ for $k = 1,\dots,n-1$. Let $\pi : R^n \to M$ be the orthogonal projection. If $\zeta_1 \leq \dots \leq \zeta_{n-1}$ are the principal semiaxes of $\pi(D)$, then $\xi_k \leq \zeta_k \leq \xi_{k+1}$ for $k = 1,\dots,n-1$.

This is a well-known geometrical fact.

(3.4) LEMMA. Let P be an arbitrary n-dimensional rectangular parallelepiped in R^n circumscribed on D. Then

$$\operatorname{diam} P = 2(\xi_1^2 + \dots + \xi_n^2)^{1/2}.$$

Proof. Let $\Phi : R^n \to R^n$ be a linear operator such that $\Phi(B_n) = D$. We can choose an orthonormal basis $(e_k)_{k=1}^n$ in R^n such that P has the form

$$\{ \sum_{k=1}^{n} t_k e_k : |t_k| \leq s_k \quad \text{for} \quad k = 1,\dots,n\}$$

for some coefficients $s_1,\dots,s_n > 0$. For each $k = 1,\dots,n$, one has

$$s_k = \sup_{u \in B_n} (\Phi u, e_k) = \sup_{u \in B_n} (u, \Phi^* e_k) = \|\Phi^* e_k\| .$$

Hence

$$\frac{1}{2} \operatorname{diam} P = (\sum_{k=1}^{n} s_k^2)^{1/2} = (\sum_{k=1}^{n} \|\Phi^* e_k\|^2)^{1/2} .$$

Now, it remains to observe that the right side is equal to the Hilbert-Schmidt norm of Φ, i.e. to $(\xi_1^2 + \dots + \xi_n^2)^{1/2}$. ▪

(3.5) LEMMA. Let M be an $(n - 1)$-dimensional affine subspace of R^n with

(1) $\quad M \cap B_n \neq \emptyset .$

Let us suppose that

(2) $\xi_1^{-2} + \ldots \xi_n^{-2} \leq 1.$

Then $D \cap M$ is an $(n-1)$-dimensional ellipsoid; denoting its principal semiaxes by $\eta_1,\ldots,\eta_{n-1}$, one has

(3) $\eta_1^{-2} + \ldots + \eta_{n-1}^{-2} \leq 1.$

Proof. That $D \cap M$ is an $(n-1)$-dimensional ellipsoid and not an empty set follows from (1) because (2) implies that int D contains B_n. Let u be the centre of $D \cap M$ and M_o the linear subspace $M - u$. We may assume that $\eta_1 \leq \ldots \leq \eta_{n-1}$. Then

(4) $\xi_k^{-1} = d_k(B_n, D)$ $(k = 1,\ldots,n),$

(5) $\eta_k^{-1} = d_k(B_n \cap M_o, (D \cap M) - u)$ $(k = 1,\ldots,n-1).$

Let $\Phi : R^n \to R^n$ be a linear operator with $\Phi(D) = B_n$. Denote $E = \Phi(B_n)$, $N = \Phi(M)$, $N_o = \Phi(M_o)$ and $w = \Phi u$. Let $\zeta_1 \geq \ldots \geq \zeta_n$ be the principal semiaxes of E. Then $\zeta_k = \xi_k^{-1}$ for $k = 1,\ldots,n$ and, by (2),

(6) $\zeta_1^2 + \ldots + \zeta_n^2 \leq 1.$

Let P be some n-dimensional rectangular parallelepiped circumscribed on E with the property that one of its $(n-1)$-dimensional faces is parallel to N. Let $\pi : R^n \to N_o$ be the orthogonal projection. From (6) and (3.4) we get $P \subset B_n$, and (1) means that $N \cap E \neq \emptyset$. Hence it easily follows that

(7) $\pi(P) \subset (B_n \cap N) - w.$

It is clear that $\pi(P)$ is an $(n-1)$-dimensional rectangular parallelepiped circumscribed on the $(n-1)$-dimensional ellipsoid $\pi(E)$. We may assume that $N_o = R^{n-1}$. Hence, by (7) and (3.4), we may write

$$\sum_{k=1}^{n-1} d_k^2(\pi(E), (B_n \cap N) - w) \leq 1.$$

This implies that

(8) $$\sum_{k=1}^{n-1} d_k^2(E \cap N_o, (B_n \cap N) - w) \leq 1$$

because $E \cap N_o \subset \pi(E)$. Since, evidently,

$$d_k(E \cap N_o, (B_n \cap N) - w) = d_k(B_n \cap M_o, (D \cap M) - u)$$

for $k = 1,\dots,n-1$, from (8) and (5) we obtain (3). ∎

(3.6) LEMMA. Let M be an $(n - 1)$-dimensional subspace of R^n, $n \geq 2$. Suppose that the principal semiaxes of D satisfy the condition $\xi_1^{-2} + \dots + \xi_n^{-2} \leq 1$. Let r be a fixed number belonging to $[0,1]$ and let E be the set of all those $u \in M$ for which the intersection of D and the straight line passing through u and perpendicular to M is a segment with length not less than $2r$. Then E is an $(n - 1)$-dimensional ellipsoid in M. If $\eta_1,\dots,\eta_{n-1}$ are its principal semiaxes, then $\eta_1^{-2} + \dots \eta_{n-1}^{-2} \leq 1$.

The proof is similar to the preceding one. It is also based on (3.4). We leave the details to the reader.

(3.7) LEMMA. Let L be a lattice in R^n with $L \cap D = \{0\}$. Then we can find a basis $(u_k)_{k=1}^n$ of L such that

$$d(u_k, \operatorname{span} \{u_i\}_{i<k}) \geq k^{-1}(\xi_1 \dots \xi_k)^{1/k}$$

for $k = 1,\dots,n$ (see (2.1)).

Proof. By (3.2), there is some $w \in L^*$ such that

$$0 < \|w\| \leq n(\xi_1 \dots \xi_n)^{-1/n}.$$

The set $\{(u,w) : u \in L\}$ is a non-zero subgroup of Z, therefore it has the form pZ for a certain $p = 1,2,\dots$. Choose some $u_n \in L$ with $(u_n,w) = p$ and let M be the orthogonal complement of w. It is clear that $L = Zu_n \oplus (L \cap M)$, and that

$$d(u_n,M) = p\|w\|^{-1} \geq n^{-1}(\xi_1 \dots \xi_n)^{1/n}.$$

We may assume that $M = R^{n-1}$. Then $L_{n-1} := L \cap R^{n-1}$ is a lattice in R^{n-1} and $D_{n-1} := D \cap R^{n-1}$ is an $(n - 1)$-dimensional ellipsoid in R^{n-1} with $L_{n-1} \cap D_{n-1} = \{0\}$. Let $\eta_1 \leq \dots \leq \eta_{n-1}$ be the principal semiaxes of D_{n-1}. By repeating the above argument, we can find some $u_{n-1} \in L_{n-1}$ and some $(n - 2)$-dimensional subspace N of R^{n-1}, such that $L_{n-1} = Zu_{n-1} \oplus (L_{n-1} \cap N)$ and

$$d(u_{n-1},N) \geq (n-1)^{-1}(\eta_1 \dots \eta_{n-1})^{1/(n-1)}.$$

From (3.3) we get $\eta_k \geq \xi_k$ for $k = 1,\dots,n-1$. Hence

$$d(u_{n-1},N) \geq (n-1)^{-1}(\xi_1 \dots \xi_{n-1})^{1/(n-1)}.$$

Next, we may assume that $N = R^{n-2}$, and so on. After n steps we shall find generators $u_n, u_{n-1}, \dots, u_1$ of L, with the desired properties. ▪

(3.8) LEMMA. Let $a \in R^n$ and let K be a subgroup of R^n with $K \cap D = \{0\}$ and $d(a,K) \geq \frac{1}{4}$. Then there exists a linear functional f on R^n such that $f(K) \subset [\frac{1}{4},\frac{1}{4}] + Z$ and

$$\|f\| \leq 1 + [\sum_{k=1}^{n} k^2(\xi_1 \dots \xi_k)^{-2/k}]^{1/2}. \tag{1}$$

Proof. By adding, if necessary, several sufficiently distant generators, we can find a lattice L in R^n with $K \subset L$, $L \cap D = \{0\}$ and $d(a,L) = d(a,K)$. So, we may assume that K is a lattice itself. Due to (3.7), we can find a basis $(u_k)_{k=1}^n$ of K such that

$$d(u_k, \operatorname{span} \{u_i\}_{i<k}) \geq k^{-1}(\xi_1 \dots \xi_k)^{1/k}$$

for $k = 1,\dots,n$. Let $w_1,\dots,w_n$ be the Gram-Schmidt orthonormalization of the system $u_1,\dots,u_n$. We may write

$$u_k = c_{k1}w_1 + \dots + c_{kk}w_k \qquad (k = 1,\dots,n)$$

for some coefficients c_{kl}. Since $c_{kk} > 0$ for every k, we have

$$c_{kk} \geq k^{-1}(\xi_1 \dots \xi_k)^{1/k} \qquad (k = 1,\dots,n). \tag{2}$$

We may write $a = a_1w_1 + \dots + a_nw_n$ for some coefficients a_k. Take $p_n \in Z$ such that

$$|a_n - p_nc_{nn}| \leq \tfrac{1}{2}c_{nn}.$$

Next, take $p_{n-1} \in Z$ such that

$$|a_{n-1} - p_n c_{n,n-1} - p_{n-1} c_{n-1,n-1}| \le \tfrac{1}{2} c_{n-1,n-1}.$$

Having found $p_n, p_{n-1}, \dots, p_k$, take $p_{k-1} \in Z$ such that

$$|a_{k-1} - \sum_{i=k-1}^{n} p_i c_{i,k-1}| \le \tfrac{1}{2} c_{k-1,k-1},$$

and so on. After n steps we shall obtain some integers $p_n, p_{n-1}, \dots, p_1$. Set $a_o = p_1 u_1 + \dots + p_n u_n$ and write $a' = a - a_o = a_1' w_1 + \dots + a_n' w_n$. Then, clearly,

(3) $$|a_k'| \le \tfrac{1}{2} c_{kk} \qquad (k = 1,\dots,n).$$

Since $a_o \in K$ and $d(a,K) \ge \frac{1}{4}$, we have $d(a',K) \ge \frac{1}{4}$, whence $\|a'\| \ge \frac{1}{4}$. Therefore we can find a functional h on R^n with $h(a') \in [\frac{1}{4}, \frac{3}{4}]$ and $\|h\| \le 1$. Set $h_k = h(w_k)$ for $k = 1,\dots,n$. We shall construct inductively a sequence $f_1, \dots, f_n$ of real coefficients such that

(4) $$|f_k - h_k| \le c_{kk}^{-1} \qquad (k = 1,\dots,n),$$

(5) $$c_{k1} f_1 + \dots + c_{kk} f_k \in Z \quad (k = 1,\dots,n),$$

(6) $$a_1' f_1 + \dots + a_n' f_n \in [\tfrac{1}{4}, \tfrac{3}{4}].$$

Put $k = 1$ in (3). Then it is not hard to see that we can find a coefficient f_1 such that $c_{11} f_1 \in Z$, $|f_1 - h_1| \le c_{11}^{-1}$ and

$$a_1' f_1 + \sum_{k=2}^{n} a_k' h_k \in [\tfrac{1}{4}, \tfrac{3}{4}].$$

Having constructed $f_1, \dots, f_{k-1}$, we can find, by (3), a coefficient f_k such that $|f_k - h_k| \le c_{kk}^{-1}$, $c_{k1} f_1 + \dots + c_{kk} f_k \in Z$ and

$$a_1' f_1 + \dots + a_k' f_k + \sum_{i=k+1}^{n} a_i' h_i \in [\tfrac{1}{4}, \tfrac{3}{4}].$$

After n steps we shall obtain $f_1, \dots, f_n$ satisfying (4) - (6).

Consider the linear functional f on R^n given by $f(w_k) = f_k$, $k = 1,\dots,n$. By (5), we have $f(u_k) \in Z$ for every k, whence $f(K) \subset Z$.

From (6) we get $f(a') \in [\frac{1}{4},\frac{3}{4}]$, so that

$$f(a) = f(a' + a_o) = f(a') + f(a_o) \in [\tfrac{1}{4},\tfrac{3}{4}] + Z.$$

Finally, (4) and (2) yield the estimate

$$\|f\| \le \|h\| + \|f - h\| \le \|h\| + [\sum_{k=1}^{n} |f_k - h_k|^2]^{1/2}$$

$$\le 1 + [\sum_{k=1}^{n} c_{kk}^{-2}]^{1/2} \le 1 + [\sum_{k=1}^{n} k^2 (\xi_1 \dots \xi_k)^{-2/k}]^{1/2}. \quad ■$$

(3.9) LEMMA. Let $a \in R^n$ and let K be a subgroup of R^n with $K \cap (a + D) = \emptyset$. Suppose that $\xi_1^{-2} + \dots + \xi_n^{-2} \le 1$. Then there exists an orthogonal projection $\pi : R^n \to R^n$ with $1 \le \dim \pi(R^n) \le n$, such that $\pi(K) \cap 2B_n = \{0\}$ and $d(\pi(a),\pi(K)) \ge 1$.

Proof. We shall apply induction on n. For $n = 1$, the lemma is trivial. Suppose that it is true for the space R^{n-1}.

If $K \cap 2B_n = \{0\}$, we take $\pi = id$. So, assume that there is some $u \in K$ with $0 < \|u\| \le 2$. Let M be the orthogonal complement of u and π_M the orthogonal projection onto M. Let E be the set of all those points $v \in M$ for which the intersection of $a + D$ with the straight line passing through v and orthogonal to M is a segment with length not less than $\|u\|$. It is not hard to see that $\pi_M(K) \cap E = \emptyset$. From (3.6) it follows that E is an ellipsoid with principal semiaxes $\eta_1,\dots,\eta_{n-1}$ such that $\eta_1^{-2} + \dots + \eta_{n-1}^{-2} \le 1$. Therefore, by our inductive assumption, there is an orthogonal projection $\sigma : M \to M$ with $1 \le \dim \sigma(M) \le n - 1$, such that $\sigma(\pi_M(K)) \cap 2(M \cap B_n) = \{0\}$ and $d(\sigma\pi_M(a),\sigma\pi_M(K)) \ge 1$. So, we may take $\pi = \sigma\pi_M$. ■

(3.10) PROPOSITION. If $a_1,a_2,\dots$ is a sequence of non-negative numbers not all zero, then

$$\sum_{n=1}^{\infty} (a_1 a_2 \dots a_n)^{1/n} < e \sum_{n=1}^{\infty} a_n.$$

For the proof, see [35], 9.12.

(3.11) LEMMA. Let $a \in R^n$ and let K be a subgroup of R^n with $K \cap (a + D) = \emptyset$. Suppose that

(1) $$\sum_{k=1}^{n} k\xi_k^{-1} \le 1.$$

Then there exists a linear functional f on R^n with $\|f\| < 4$, $f(K) \subset Z$ and $f(a) \in [\frac{1}{4},\frac{3}{4}] + Z$.

Proof. We may assume that

$$D = \{(x_1,\ldots,x_n) \in R^n : \sum_{k=1}^{n} \xi_k^{-2} x_k^2 \le 1\}.$$

Denote $\eta_k = (k\xi_k)^{1/2}$ for $k = 1,\ldots,n$, and

$$E = \{(x_1,\ldots,x_n) \in R^n : \sum_{k=1}^{n} \eta_k^{-2} x_k^2 \le 1\}.$$

From (1) we get $\sum_{k=1}^{n} \xi_k^{-2}\eta_k^2 \le 1$. Thus, by (3.9), there exists an orthogonal projection $\pi : R^n \to R^n$ with $m := \dim \pi(R^n) = 1,\ldots,n$, such that $\pi(K) \cap 2E = \{0\}$ and $(\pi(a) + E) \cap \pi(K) = \emptyset$. Hence, by (1),

$$d(\pi(a),\pi(K)) \ge \eta_1 = \xi_1^{1/2} > 1 > \frac{1}{4}.$$

Let $\zeta_1 \le \ldots \le \zeta_m$ be the principal semiaxes of the ellipsoid $\pi(R^n) \cap 2E$. By (3.8), there exists a linear functional h on R^n with $h(\pi(K)) \subset Z$, $h(\pi(a)) \in [\frac{1}{4},\frac{3}{4}] + Z$ and

$$\|h\| \le 1 + [\sum_{k=1}^{m} k^2(\zeta_1 \ldots \zeta_k)^{-2/k}]^{1/2}.$$

Set $f = h\pi$. From (3.3) it follows that $\zeta_k \ge 2\eta_k$ for $k = 1,\ldots,m$. Hence, applying (1), (3.10) and the inequality $k(k!)^{-1/k} < e$, we get

$$\|f\| = \|h\pi\| \le \|h\|\cdot\|\pi\| = \|h\| \le 1 + \frac{1}{2}[\sum_{k=1}^{m} (\eta_1 \ldots \eta_k)^{-2/k}]^{1/2}$$

$$= 1 + \frac{1}{2}[\sum_{k=1}^{m} k^2(k!)^{-2/k}(1^2\eta_1^{-2} \ldots k^2\eta_k^{-2})^{1/k}]^{1/2}$$

$$< 1 + \frac{1}{2} e^{3/2}[\sum_{k=1}^{m} k^2\eta_k^{-2}]^{1/2} = 1 + \frac{1}{2}[\sum_{k=1}^{m} k\xi_k^{-1}]^{1/2}$$

$$< 1 + \frac{1}{2} e^{3/2}[\sum_{k=1}^{n} k\xi_k^{-1}]^{1/2} \le 1 + \frac{1}{2} e^{3/2} < 4. \quad \blacksquare$$

(3.12) **LEMMA.** Let $w_1,\dots,w_n$ be the Gram-Schmidt orthogonalization of some system $u_1,\dots,u_n \in D$. Then $\sum_{k=1}^{n} \|w_k\|^2 \leq \sum_{k=1}^{n} \xi_k^2$.

This is a direct consequence of (3.4).

(3.13) **LEMMA.** Let E be an ellipsoid in R^n with centre at zero and principal semiaxis $\eta_1,\dots,\eta_n$ such that

(1) $\eta_1^2 + \dots + \eta_n^2 < \frac{1}{4}$.

Let K be a subgroup of R^n and χ a character of K, such that

(2) $|\chi(K \cap B_n)| < \frac{1}{3}$.

Then we can find a closed subgroup K' of R^n with $K \subset K'$ and a character χ' of K' with $\chi'_{|K} = \chi$, such that all non-zero components of K' are disjoint from E, and

(3) $|\chi'(K' \cap \frac{1}{2}B_n)| \leq |\chi(K \cap B_n)|$.

Proof. By (2) and (1.2), we have

$$|\chi(K \cap \tfrac{1}{2}B_n)| \leq \tfrac{1}{2}|\chi(K \cap B_n)| < \tfrac{1}{6} < \tfrac{1}{4}.$$

Therefore, χ is a continuous character of K due to (1.4). The continuous extension $\overline{\chi}$ of χ onto $\overline{K}$ is a continuous character of $\overline{K}$. Denote $B_n^o = \operatorname{int} B_n$. From (2) we obtain

(4) $|\overline{\chi}(K \cap B_n^o)| \leq |\chi(K \cap B_n)| < \frac{1}{3}$

because B_n^o is open. If all non-zero components of $\overline{K}$ are disjoint from E, we may take $K' = \overline{K}$ and $\chi' = \overline{\chi}$. So, assume that some non-zero component M of $\overline{K}$ has common points with E. Let M_o be the zero component of $\overline{K}$.

Let $w_1 \in M$ be the vector orthogonal to M_o. We may assume that $tw_1 \notin \overline{K}$ for $t \in (0,1)$. It is easy to verify that the formula

$$\chi_1(u + tw_1) = \rho[\overline{\chi}(u) + t\overline{\chi}(w_1)] \qquad (u \in \overline{K};\ t \in R)$$

defines a character χ_1 of the group $K_1 := \overline{K} + Rw_1$. Notice that K_1 is closed, being the vector sum of the closed group $\overline{K}$ and of the compact interval $[0,w_1]$. Obviously, $\chi_{1|K} = \chi$. Denote

$$A_1 = \{u \in B_n^o : u + w_1, u - w_1 \in B_n^o\}.$$

It is clear that A_1 is symmetric and convex. We shall prove that

$$(5) \qquad |\chi_1(K_1 \cap A_1)| \le |\bar{\chi}(K \cap B_n^o)|.$$

So, take any $v \in K_1 \cap A_1$. We may write $v = (1 - t)u_1 + tu_2$ for some $u_1, u_2 \in K \cap B_n^o$ with $u_2 - u_1 = w_1$ and for some $t \in [0,1)$. It is not difficult to see that u_1, u_2 and w_1 all belong to $K \cap B_n^o$. Hence

$$(6) \qquad |\bar{\chi}(u_1)|, |\bar{\chi}(u_2)|, |\bar{\chi}(w_1)| \le |\bar{\chi}(K \cap B_n^o)|.$$

In virtue of (4) and (1.1), we have

$$\bar{\chi}(u_2) = \bar{\chi}(u_1) + \bar{\chi}(w_1).$$

Hence, by (6),

$$\begin{aligned}|\chi_1(v)| &= |\chi_1(u_1 + tw_1)| = |\rho[\bar{\chi}(u_1) + t\bar{\chi}(w_1)]| \\ &= |\rho[(1-t)\bar{\chi}(u_1) + t\bar{\chi}(u_2)]| \le |(1 - t)\bar{\chi}(u_1) + t\bar{\chi}(u_2)| \\ &\le |\bar{\chi}(K \cap B_n^o)|.\end{aligned}$$

This proves (5).

If all non-zero components of K_1 are disjoint from E, then we stop. In the other case, we can repeat the above argument. We shall find some vector w_2 orthogonal to $M_o + Rw_1$ such that $w_2 + M_o + Rw_1$ is a non-zero component of K_1 having common points with E. We shall also obtain an extension χ_2 of χ_1 onto the closed subgroup $K_2 = K_1 + Rw_2$, such that $|\chi_2(K_2 \cap A_2)| \le |\chi_1(K_1 \cap A_1)|$ where

$$A_2 = \{u \in A_1 : u + w_2, u - w_2 \in A_1\}.$$

Since A_1 was symmetric and convex, so is A_2.

Then we proceed by induction. After several steps we shall obtain some vectors $w_1, w_2, \dots, w_p$ such that the group $K_p = \bar{K} + Rw_1 + \dots + Rw_p$ is closed and all its non-zero components are disjoint from E (it may happen that $K_p = R^n$). The system $w_1, \dots, w_p$ is obtained by the Gram-

-Schmidt orthogonalization of some system of vectors belonging to E. We shall also obtain a sequence $A_1 \supset A_2 \supset \dots \supset A_p$ of symmetric, convex subsets of R^n such that

$$A_{k+1} = \{u \in A_k : u + w_{k+1}, u - w_{k+1} \in A_k\} \qquad (k = 1,\dots,p-1)$$

and a character χ_p of K_p with $\chi_{p|K} = \chi$ and $|\chi_p(K_p \cap A_p)| \le |\bar{\chi}(K \cap B_n^o)|$.

Set $K' = K_p$ and $\chi' = \chi_p$. To verify (3), it is enough to show that $\frac{1}{2}B_n \subset A_p$. So, take an arbitrary $u \in \frac{1}{2}B_n$. In order to prove that $u \in A_p$, we have to show that $u + \varepsilon_p w_p \in A_{p-1}$ for $\varepsilon_p = \pm 1$. Next, to prove that $u + \varepsilon_p w_p \in A_{p-1}$, we have to show that $u + \varepsilon_p w_p + \varepsilon_{p-1} w_{p-1} \in A_{p-2}$ for $\varepsilon_{p-1} = \pm 1$, and so on. Thus, we should prove that $u + \varepsilon_p w_p + \dots + \varepsilon_1 w_1 \in B_n^o$ for all systems of signs $\varepsilon_p, \dots, \varepsilon_1 = \pm 1$. From (3.12) we have

$$\|w_1\|^2 + \dots + \|w_p\|^2 \le \eta_1^2 + \dots + \eta_n^2.$$

Hence, by (1),

$$\|u + \varepsilon_p w_p + \dots + \varepsilon_1 w_1\| \le \|u\| + \|\varepsilon_p w_p + \dots + \varepsilon_1 w_1\|$$

$$\le \frac{1}{2} + (\|w_p\|^2 + \dots + \|w_1\|^2)^{1/2} < \frac{1}{2} + \frac{1}{2} = 1. \quad \blacksquare$$

(3.14) LEMMA. Let K be a subgroup of R^n with $K \cap D = \{0\}$ and let χ be a character of K. Then we can find a linear functional f on R^n with $\rho f_{|K} = \chi$ and

$$(1) \qquad \|f\| \le \frac{1}{2} e^{3/2} \Big[\sum_{k=1}^{n} k^2 \xi_k^{-2}\Big]^{1/2}.$$

Proof. We may assume K to be a lattice in R^n. According to (3.7), we can find a basis $u_1,\dots,u_n$ of K such that

$$d(u_k, \operatorname{span}\{u_i\}_{i=1}^{k-1}) \ge k^{-1}(\xi_1 \dots \xi_k)^{1/k}$$

for $k = 1,\dots,n$. Let $e_1,\dots,e_n$ be the Gram-Schmidt orthogonalization of the system $u_1,\dots,u_n$. Then

$$u_k = a_{k1}e_1 + \dots + a_{kk}e_k \qquad (k = 1,\dots,n)$$

for some coefficients a_{kl} with

$$|a_{kk}| \geq k^{-1}(\xi_1 \ldots \xi_k)^{1/k} \qquad (k = 1,\ldots,n).$$

Now, we can find, in succession, coefficients $f_n, f_{n-1}, \ldots, f_1$ such that

$$\rho(a_{k1}f_1 + \ldots + a_{kk}f_k) = \chi(u_k) \qquad (k = 1,\ldots,n).$$

$$|f_k| \leq |2a_{kk}|^{-1} \qquad (k = 1,\ldots,n).$$

By taking $f(e_k) = f_k$ for $k = 1,\ldots,n,$ we obtain some linear functional f on R^n with $\rho f(u_k) = \chi(u_k)$ for $k = 1,\ldots,n.$ Hence $\rho f_{|K} = \chi$ because $(u_k)_{k=1}^n$ was a basis of K. Finally, by (3.10) and the inequality $k(k!)^{-1/k} < e,$ we get

$$\|f\|^2 = \sum_{k=1}^n f_k^2 \leq \sum_{k=1}^n (2a_{kk})^{-2} \leq \frac{1}{4} \sum_{k=1}^n k^2(\xi_1 \ldots \xi_k)^{-2/k}$$

$$= \frac{1}{4} \sum_{k=1}^n k^2(1^2\xi_1^{-2} \cdot 2^2\xi_2^{-2} \cdot \ldots \cdot k^2\xi_k^{-2})^{1/k}(k!)^{-2/k}$$

$$< \frac{1}{4} e^2 \sum_{k=1}^n (1^2\xi_1^{-2} \cdot \ldots \cdot k^2\xi_k^{-2})^{1/k} < \frac{1}{4} e^3 \sum_{k=1}^n k^2\xi_k^{-2}. \quad \blacksquare$$

(3.15) LEMMA. Let K be a subgroup of R^n and χ a character of K with $|\chi(K \cap D)| \leq \frac{1}{4}$. Then there exists a linear functional f on R^n with $\rho f_{|K} = \chi$ and $\|f\| \leq 5 \sum_{k=1}^n k\xi_k^{-1}$.

Proof. Without loss of generality we may assume that

$$(1) \qquad \sum_{k=1}^n k\,\xi_k^{-1} = 1.$$

Next, we may assume that

$$D = \{(x_1,\ldots,x_n) \in R^n : \sum_{k=1}^n x_k^2\xi_k^{-2} \leq 1\}.$$

Define

$$E = \{(x_1,\ldots,x_n) \in R^n : \sum_{k=1}^n k^{-1}x^2\xi_k^{-1} \leq \ ^2\}$$

where $\vartheta = 0.48$. By (1), we have

$$\sum_{k=1}^{n} d_k^2(E,D) = \vartheta^2 \sum_{k=1}^{n} k\xi_k^{-1} < \frac{1}{4}.$$

So, by (3.13), there exist a closed subgroup K' of R^n such that all its non-zero components are disjoint from E and a character χ' of K' with $\chi'_{|K} = \chi$ and

(2) $\quad |\chi'(K \cap \frac{1}{2}D)| \leq |\chi(K \cap D)|.$

Let K'_o be the zero component of K'. By (2.3), there is a linear functional h on K'_o with $\rho h = \chi'_{|K'_o}$. From (2) it follows that $|\rho h(u)| \leq \frac{1}{4}$ for $u \in K'_o \cap \frac{1}{2}D$. Hence $|h(u)| \leq \frac{1}{4}$ for $u \in K'_o \cap \frac{1}{2}D$. From (1) we get $B_n \subset D$, which implies that

(3) $\quad \|h\| \leq \frac{1}{2}.$

Let M be the orthogonal complement of K'_o in R^n and let $r = \dim M$. Denote $N = K' \cap M$. According to (3.1), we have $K' = K'_o \oplus N$. Let E_M be the orthogonal projection of E onto M. Let $\eta_1 \leq \ldots \leq \eta_n$ and $\zeta_1 \leq \ldots \leq \zeta_r$ be the principal semiaxes of E and E_M, respectively. Applying (3.3) several times, we see that $\zeta_k \geq \eta_k$ for $k = 1,\ldots,r$. We have $\eta_k = \vartheta k^{1/2}\xi_k^{1/2}$ for $k = 1,\ldots,n$; therefore

(4) $\quad \zeta_k \geq \vartheta k^{1/2}\xi_k^{1/2} \qquad (k = 1,\ldots,r).$

Since non-zero components of K' are disjoint from E, it follows that $N \cap E_M = \{0\}$. In virtue of (3.14), there is a linear functional $h^\perp$ on M with $\rho h^\perp_{|N} = \chi'_{|N}$ and

$$\|h^\perp\| \leq \frac{1}{2}e^{3/2}\Big[\sum_{k=1}^{r} k^2\zeta_k^{-2}\Big]^{1/2}.$$

Applying (4) and (1), we obtain

(5) $\quad \|h^\perp\| \leq \frac{1}{2\vartheta}\, e^{3/2}.$

Let π and $\pi^\perp$ be the orthogonal projections of R^n onto K'_o and M, respectively. Set $f = h\pi + h^\perp\pi^\perp$. An easy verification shows that $\rho f_{|K} = \chi$. From (3) and (5) we obtain

$$\|f\| \leq (\|h\|^2 + \|h^\perp\|^2)^{1/2} \leq (\tfrac{1}{4} + \frac{1}{4\theta^2} e^3)^{1/2} < 5. \quad \blacksquare$$

(3.16) LEMMA. Let U be a symmetric, radial subset of R^n and let L be a lattice in R^n such that $gp(L \cap U) = L$. Then we can find generators $u_1,\ldots,u_n$ of L such that $u_k \in U + \text{span}\,\{u_i\}_{i<k}$ for $k = 1,\ldots,n$ (see (2.1)).

Proof. Choose some $u \in L \cap U$. Let u_1 be one of the generators of the group $L \cap Ru$. We have $u_1 = tu$ for a certain $t \in [-1,1]$, which implies that $u_1 \in U$ because U is symmetric and radial.

Now, suppose that, for a certain $m = 1,\ldots,n-1$, we have found some linearly independent vectors $u_1,\ldots,u_m \in L$ such that

$$u_k \in U + \text{span}\,\{u_i\}_{i<k} \qquad (k = 1,\ldots,m),$$

$$gp(\{u_k\}_{k=1}^m) = L \cap \text{span}\,\{u_k\}_{k=1}^m.$$

Let us write $M = \text{span}\,\{u_k\}_{k=1}^m$. If $L \cap (U + M) \subset M$, then

$$L = gp\,(L \cap U) \subset gp\,(L \cap (U + M)) \subset gp\,(M) = M,$$

whence $\dim \text{span}\, L \leq \dim M = m < n$, which is impossible. So, we can find some $v \in L \cap (U + M)$ with $v \notin M$. Let u_{m+1} be one of the elements of the set $[L \cap (M + Rv)] \setminus M$ which are nearest to M. We have $u_{m+1} \in L \cap (U + M)$ because U is radial and symmetric. It is not difficult to observe that $gp\,(\{u_k\}_{k=1}^{m+1}) = L \cap (M + Ru_{m+1})$. If $m + 1 = n$, we are through. If not, then we can repeat the above argument to obtain some vector u_{m+2}, and so on. After a finite number of steps we shall obtain vectors $u_1,\ldots,u_n$ with the desired properties. $\blacksquare$

(3.17) LEMMA. Let L be a lattice in R^n with $gp\,(L \cap D) = L$. Suppose that $\xi_1^2 + \ldots + \xi_n^2 \leq 1$. Then we can find a rectangular parallelepiped $P \subset \frac{1}{2}B_n$ such that $\{u + P\}_{u \in L}$ is a disjoint covering of R^n.

Proof. Due to (3.16), we can find generators $u_1,\ldots,u_n$ of L such that $u_k \in D + \text{span}\,\{u_i\}_{i<k}$ for $k = 1,\ldots,n$. So, there are some $v_1,\ldots,v_n \in D$ such that $u_k \in v_k + \text{span}\,\{u_i\}_{i<k}$ for $k = 1,\ldots,n$. Let $w_1,\ldots,w_n$ be the Gram-Schmidt orthogonalization of the system $u_1,\ldots,u_n$.

Then $w_1,\ldots,w_n$ is the Gram-Schmidt orthogonalization of $v_1,\ldots,v_n$; so, by (3.12),

$$(1) \qquad \|w_1\|^2 + \ldots + \|w_n\|^2 \le \xi_1^1 \ldots \xi_n^2 \le 1.$$

Set

$$P = \{t_1w_1 + \ldots + t_nw_n : -\tfrac{1}{2} \le t_1,\ldots,t_n < \tfrac{1}{2}\}.$$

It is not hard to see that $\{u + P\}_{u\in L}$ is a disjoint covering of R^n. From (1) we get $P \subset \frac{1}{2} B_n$. ∎

(3.18) COROLLARY. Let p,q be two pre-Hilbert seminorms on a vector space E, with

$$(1) \qquad \sum_{k=1}^{\infty} d_k^2(B_p,B_q) \le 1.$$

Let K be a subgroup of E such that

$$(2) \qquad K = \text{gp}\,(K \cap B_p).$$

Then $\text{span}\, K \subset K + \frac{1}{2}B_q$.

Proof. Choose any $u \in \text{span}\, K$. In view of (2), we may write $u = \alpha_1u_1 + \ldots + \alpha_nu_n$ for some $u_1,\ldots,u_n \in K \cap B_p$ and some coefficients $\alpha_1,\ldots,\alpha_n$. We may assume $u_1,\ldots,u_n$ to be linearly independent.

Denote $M = \text{span}\,\{u_k\}_{k=1}^n$ and $L = \text{gp}\,\{u_k\}_{k=1}^n$. Then L is a lattice in M. From (1) and (2.13) we obtain

$$\sum_{k=1}^{n} d_k^2(B_p \cap M, B_q \cap M) \le 1.$$

So, according to (3.17), the family $\{v + \frac{1}{2}(B_q \cap M)\}_{v\in L}$ is a covering of M. Hence $u \in M \subset L + \frac{1}{2}(B_q \cap M) \subset K + \frac{1}{2}B_q$. ∎

(3.19) LEMMA. Let K be a subgroup of R^n with $\text{span}\, K = R^n$ and $\text{gp}\,(K \cap D) = K$. Suppose that $\xi_1^2 \ldots \xi_n^2 \le \frac{1}{4}$. Then $\frac{1}{2}B_n \subset \text{conv}\,(K\cap B_n)$.

Proof. Without loss of generality we may assume that K is a lattice in R^n. It follows from (3.18) that

$$(1) \qquad K + \tfrac{1}{2}B_n = R^n.$$

Suppose that there is some $u \in (\frac{1}{2}B_n) \setminus \mathrm{conv}\,(K \cap B_n)$. Since K is a lattice, $\mathrm{conv}\,(K \cap B_n)$ is compact. So, there is some $f \in (R^n)^*$ with $f(u) > 1$, such that $f < 1$ on $K \cap B_n$. We have $\|f\| > 2$ because $\|u\| \le \frac{1}{2}$. Choose $w \in R^n$ such that $\|w\| = \frac{3}{4}$ and $f(w) = \|f\|\cdot\|w\|$. By (1), there is some $v \in K$ with $\|w - v\| \le \frac{1}{4}$. Then $\|v\| \le \|w - v\| + \|w\| \le 1$, i.e. $v \in K \cap B_n$, whence $f(v) < 1$. Thus

$$\|f\|\cdot\|w\| - f(v) = |f(w) - f(v)| \le \|f\|\cdot\|w - v\|$$

and, consequently,

$$\frac{1}{4} < \|w\| - \|f\|^{-1}\cdot f(v) \le \|w - v\| \le \frac{1}{4},$$

which is impossible. ▪

(3.20) COROLLARY. Let p,q be two pre-Hilbert seminorms on a vector space E, with

(1) $$\sum_{k=1}^{n} d_k^2(B_p,B_q) \le \frac{1}{4}.$$

Then, for any subgroup K of E, we have

$$d_k(\mathrm{conv}\,(K \cap B_p), \mathrm{conv}\,(K \cap B_q)) \le 2d_k(B_p,B_q) \qquad (k = 1,2,\ldots).$$

Proof. Set $M = \mathrm{span}\,(K \cap B_p)$. We shall prove that

(2) $$M \cap \tfrac{1}{2}B_q \subset \mathrm{conv}\,(K \cap B_q).$$

So, choose any $u \in M \cap \frac{1}{2}B_q$. We shall find some $w_1,\ldots,w_n \in K \cap B_p$ such that $u \in N : \mathrm{span}\,\{w_i\}_{i=1}^n$. Let $L = \mathrm{gp}\,\{w_i\}_{i=1}^n$ and let r,s be the restrictions to N of p,q, respectively. We have $w_1,\ldots,w_n \in B_r$ and $u \in \frac{1}{2}B_s$. From (1) and (2.13) it follows that

$$\sum_{k=1}^{\infty} d_k^2(B_r,B_s) = \sum_{k=1}^{\infty} d_k^2(B_p \cap M, B_q \cap M) \le \sum_{k=1}^{\infty} d_k^2(B_p,B_q) \le \frac{1}{4}.$$

Hence $u \in \mathrm{conv}\,(L \cap B_s)$ according to (3.19). This proves (2). From (2) and (2.13) we get

$$d_k\,(\mathrm{conv}\,(K \cap B_p), \mathrm{conv}\,(K \cap B_q)) \le d_k(M \cap B_p, M \cap \tfrac{1}{2}B_q)$$

$$\le d_k(B_p,\tfrac{1}{2}B_q) = 2d_k(B_p,B_q) \qquad (k = 1,2,\ldots). \quad ▪$$

Let us formulate the results of section 3 in the language of geometry of numbers. To this aim, we have to introduce some notions. Let L be a lattice in R^n and let U be a symmetric, convex body in R^n. The <u>successive</u> <u>minima</u> of L with respect to U are defined in the following way:

$$\lambda_i(L,U) = \inf \{r > 0 : \dim \operatorname{span} (L \cap rU) \geq i\} \qquad (i = 1,\dots,n).$$

The quantity

$$\mu(L,U) = \inf \{r > 0 : L + rU = R^n\}$$

is called the <u>covering</u> <u>radius</u> of L with respect to U.

(3.21) LEMMA. Let U be a symmetric convex body in R^n. For each lattice L in R^n, one has

$$[\mu(L,U)]^n \geq d(L)/\mathrm{vol}_n(U).$$

This is a standard fact; see e.g. inequality (8) in [28], Ch. XI, §1, n° 3.

Now, (3.2) can be written in the following way: for each lattice L in R^n, one has

$$\lambda_1(L^*,B_n)\lambda_1(L,D) \leq n(\xi_1 \dots \xi_n)^{-1/n}.$$

The proof of (3.2) also implies that

$$\lambda_1(L^*,B_n)\lambda_1(L,B_n) \leq 4\omega_n^{-2/n} \approx \frac{2n}{\pi e}.$$

On the other hand, Conway and Thompson proved that to each $n = 1,2,\dots$ there corresponds a lattice $L_n = L_n^*$ in R^n such that

$$\lambda_1^2(L_n,B_n) > (\tfrac{5}{3}\omega_n^{-1})^{2/n} - \frac{1}{2} \approx \frac{n}{2\pi e}$$

(see [66], Ch. II, Theorem 9.5).

In connection with these inequalities, it is worth, perhaps, noticing the following thing. Let U be a symmetric convex body in R^n and U^0 the polar body, that is,

$$U^0 = \{u \in R^n : (u,v) \leq 1 \text{ for all } v \in U\}.$$

The Bourgain-Milman inequality

$$\mathrm{vol}_n(U)\ \mathrm{vol}_n(U^0) \geq c^n \omega_n^2$$

(see [24]) implies that

$$\lambda_1(L,U)\lambda_1(L^*,U^0) \leq c_1 n$$

for each lattice L in R^n; here c and c_1 are some universal constants. On the other hand, from Siegel's mean value theorem one can deduce that there exists a universal constant c_2 such that to each symmetric convex body U in R^n there corresponds a lattice L with

$$\lambda_1(L,U)\lambda_1(L^*,U^0) \geq c_2 n;$$

the proof will be given somewhere else.

Let L be a lattice in R^n and $a_1,\ldots,a_n$ some fixed basis of L. For each $x = (x_1,\ldots,x_n) \in R^n$, let F_x be the set of all linear functionals f on R^n such that $f(a_i) \in Z + x_i$ for $i = 1,\ldots,n$. Then it is clear that

(1) $$\mu(L^*,B_n) = \max_{x\in R^n} \min_{f\in F_x} \|f\|.$$

Lemma (3.14) says that if $L \cap D = \{0\}$, then to each $x \in R^n$ there corresponds some $f \in F_x$ with

$$\|f\| \leq \tfrac{1}{2}e^{3/2}\left[\sum_{k=1}^{n} k^2\xi_k^{-2}\right]^{1/2}.$$

Then, by (1),

$$\mu(L^*,B_n) \leq \tfrac{1}{2}e^{3/2}\left[\sum_{k=1}^{n} k^2\xi_k^{-2}\right]^{1/2}.$$

Consequently, for each lattice L in R^n, one has

$$\mu(L^*,B_n)\lambda_1(L,D) \leq \tfrac{1}{2}e^{3/2}\left[\sum_{k=1}^{n} k^2\xi_k^{-2}\right]^{1/2}.$$

In particular, if $D = B_n$, then

$$\mu(L^*,B_n)\lambda_1(L,B_n) \leq \tfrac{1}{2}e^{3/2}\left[\sum_{k=1}^{n} k^2\right]^{1/2} < 2.25n^{3/2}.$$

A somewhat better inequality

$$\mu(L^*,B_n)\lambda_1(L,B_n) \leq \tfrac{1}{2}n^{3/2}$$

was obtained in [58], Theorem (2.14). On the other hand, from (3.21) and the Minkowski-Hlawka theorem it follows that, for every n, there exists a lattice L in R^n such that

$$(2) \qquad \mu(L^*,B_n)\lambda_1(L,B_n) > \omega_n^{-2/n} \approx \frac{n}{2\pi e}.$$

The methods applied in section 3 also allow to prove the following result: for every lattice L in R^n, one has

$$\lambda_i(L,D)\lambda_{n-i+1}(L^*,B_n) \le 6 \sum_{k=1}^{n} k\,\xi_k^{-1} \qquad (i = 1,\ldots,n).$$

The detailed proof is given in [11]. In particular, taking $\xi_1 = \ldots = \xi_n = 1$, we obtain

$$\lambda_i(L,B_n)\lambda_{n-i+1}(L^*,B_n) \le 3n(n+1) \qquad (i = 1,\ldots,n).$$

This differs only by a constant factor from the bounds

$$\lambda_i(L,B_n)\lambda_{n-i+1}(L^*,B_n) \le \tfrac{1}{6}n^2 \qquad (i = 1,\ldots,n)$$

which were obtained in [58] for $n \ge 7$.

Let L be a lattice in R^n. From (3.11) it follows immediately that to each $a \in R^n$ there corresponds some $v \in L^*$ such that

$$(3) \qquad \|v\|^{-1}\rho((v,a)) \ge [8n(n+1)]^{-1}d(a,L).$$

This result was independently obtained by Hastad [36], with $[8n(n+1)]^{-1}$ replaced by $[6n^2+1]^{-1}$. It follows from (2) that the right side of (3) cannot be replaced by $cn^{-1}d(a,L)$ with c greater than πe.

Papers [58] and [36] are based on the notion of the so-called <u>Korkin-Zolotarev bases</u> (see [58]). The proofs of (3.11) and (3.14) given above are, in fact, similar to those given in [36] and [58], respectively; Korkin-Zolotarev bases occur in the proofs of (3.8) and (3.14).

(3.22) NOTE. The material of this section is taken from [5], [7] and [8]. The idea of applying (3.4) in the proofs of (3.5), (3.6) and (3.12) comes from [10]. Lemma (3.15) is a strengthening of Lemma 1.5 of [8]. Lemma (3.17) and Corollary (3.18) can be found in [15]; (3.19) and (3.20) are new.

Chapter 2

EXOTIC GROUPS

It is not hard to find abelian topological groups without (non--trivial) continuous characters; perhaps the simplest examples are the spaces L^p, $0 \leq p < 1$ (see (2.3) or [38], (23.32)). It is much more difficult to find abelian groups without non-trivial continuous unitary representations (the so-called <u>exotic groups</u>). The first example of such a group was obtained only in 1974 by W. Herer and J.P.R. Christensen [37]. In this chapter we present various constructions of abelian groups without continuous characters or unitary representations. Section 4 wears a preliminary complexion; we gather here several more special technical results on continuous representations of abelian topological groups.

4. Representations of abelian topological groups

Let X be a measure space with a positive measure μ. By $L^2_C(X,\mu)$ we denote the complex Hilbert space of all (classes of) square-integrable functions on X, with the usual norm. By $L^\infty_C(X,\mu)$ we denote the complex Banach space of all (classes of) essentially bounded functions on X, with the ess sup norm. We may treat $L^\infty_C(X,\mu)$ as an algebra of operators in $L^2_C(X,\mu)$, identifying a function belonging to $L^\infty_C(X,\mu))$ with the corresponding operator of pointwise multiplication.

If $\mu(X) < \infty$, we define $L^0_C(X,\mu)$ as the complex space of all (classes of) measurable functions on X, with the topology of convergence in measure. This topology can be defined by the F-norm

$$|f| = \int_X \min(1,|f(x)|)d\mu(x).$$

If A is a subset of C and $p = 0, 2, \infty$, then by $L^p_A(X,\mu)$ we denote the subset of $L^p_C(X,\mu)$ consisting of A-valued functions. If $X = (0,1)$, and μ is the Lebesgue measure, we write $L^p_A(0,1)$ instead of $L^p_A(X,\mu)$.

By an $L_S(X,\mu)$-<u>representation</u> of a group G we mean a representation of G in the space $L^2_C(X,\mu)$ by operators belonging to $L^0_S(X,\mu)$.

Naturally, such representations are unitary.

If Θ is a linear operator from a vector space E to the space $L^0_R(X,\mu)$, then the formula

$$(\Phi_u f)(x) = f(x)\cdot\exp\,[2\pi i(\Theta u)(x)] \qquad (u\in E;\quad f\in L^2_C(X,\mu);\quad x\in X)$$

defines an $L_S(X,\mu)$-representation Φ of E; we write $\Phi = e^{2\pi i\Theta}$.

(4.1) PROPOSITION. A linear operator Θ from a topological vector space E to $L^0_R(X,\mu)$ is continuous if and only if $e^{2\pi i\Theta}$ is a continuous representation of E.

In the proof we need the following fact:

(4.2) LEMMA. Let X be a measure space with a finite, positive measure μ. If $f\in L^0_R(X,\mu)$ and $|f(x)|\geq 1$ a.e., then we can find some $t\in(0,1)$ with

$$\mu(\{x\in X : tf(x)\in[\tfrac{1}{4},\tfrac{3}{4}] + Z\}) \geq \tfrac{2}{5}\mu(X).$$

Proof. Let λ be the Lebesgue measure on $(0,1)$. If $x\in X$ and $|f(x)|\geq 1$, then an easy argument shows that

$$\lambda(\{t\in(0,1) : tf(x)\in[\tfrac{1}{4},\tfrac{3}{4}] + Z\}) \geq \tfrac{2}{5}.$$

Hence, by the Fubini theorem, we obtain

$$\int_0^1 \mu(\{x\in X : tf(x)\in[\tfrac{1}{4},\tfrac{3}{4}] + Z\})dt$$

$$= \int_X \lambda(\{t\in(0,1) : tf(x)\in[\tfrac{1}{4},\tfrac{3}{4}] + Z\})d\mu(x) \geq \tfrac{2}{5}\mu(X).$$

This implies that the function under the first integral must assume a value not less than $\frac{2}{5}\mu(X)$ for a certain $t\in(0,1)$. ∎

Proof of (4.1). Denote $\Phi = e^{2\pi i\Theta}$. Suppose first that Θ is continuous. To prove the continuity of Φ, choose any $f\in L^2_C(0,1)$ and $\varepsilon > 0$. There is some $\delta > 0$ such that if Y is a measurable subset of X with $\mu(Y) < \delta$, then

$$\int_Y |f(x)|^2 d\mu(x) < \varepsilon.$$

Since Θ is continuous, there is some $U \in N_o(E)$ such that

$$\mu(\{x \in X : |(\Theta u)(x)| > \varepsilon\}) \leq \delta$$

for $u \in U$. Now, choose any $w,v \in E$ with $w - v \in U$ and denote

$$Y = \{x \in X : |(\Theta w)(x) - (\Theta v)(x)| > \varepsilon\}.$$

Then

$$\|\Phi_w f - \Phi_v f\|^2$$

$$= \int_X |f(x)\cdot\exp[2\pi i(\Theta w)(x)] - f(x)\cdot\exp[2\pi i(\Theta v)(x)]|^2 d\mu(x)$$

$$= \int_Y + \int_{X\setminus Y} |\exp[2\pi i(\Theta w)(x)]$$

$$- \exp[2\pi i(\Theta v)(x)]|^2\cdot|f(x)|^2 d\mu(x)$$

$$\leq 4\int_Y |f(x)|^2 d\mu(x) + (2\pi\varepsilon)^2 \int_{X\setminus Y} |f(x)|^2 d\mu(x)$$

$$\leq 4\varepsilon + 4\pi^2\varepsilon^2\|f\|^2.$$

Since $\varepsilon > 0$ was arbitrary, this proves that Φ is continuous.

Now, suppose that Θ is not continuous. Then we can find a constant $c > 0$ such that each $U \in N_o(E)$ contains a vector u with

$$\mu(\{x \in X : |(\Theta u)(x)| \geq c\}) \geq c.$$

Hence, by (4.2), each $U \in N_o(E)$ contains a vector u with

(1) $$\mu(\{x \in X : (\Theta u)(x) \in [\tfrac{1}{4},\tfrac{3}{4}] + Z\}) \geq \tfrac{2}{5}c$$

(we use twice the fact that E has a base at zero consisting of radial sets). Set $f_o \equiv 1$. From (1) we get

$$\|\Phi_u f_o - \Phi_0 f_o\|^2 = \int_X |1 - \exp[2\pi i(\Theta u)(x)]|^2 d\mu(x) \geq \tfrac{4}{5}c$$

becasue $|1 - \exp[2\pi is]|^2 \geq 2$ whenever $s \in [\tfrac{1}{4},\tfrac{3}{4}] + Z$. This means that Φ is not continuous. ∎

Observe that the representation $e^{2\pi i\Theta}$ is uniformly continuous if and only if Θ is a continuous operator from E to $L_R^\infty(0,1)$.

(4.3) PROPOSITION. Let X be a measure space with a positive measure μ and let Φ be a continuous $L_S(X,\mu)$ - representation of the group R. Then there is a unique function $\Theta \in L_R^0(X,\mu)$ with

$$(\Theta_s f)(x) = f(x)\cdot\exp\,[2\pi is\,(x)] \qquad (s \in R;\quad f \in L_C^2(X,\mu);\quad x \in X).$$

This is a consequence of Stone's theorem on continuous one-parameter groups of unitary operators (see e.g. [47], Theorem 5.6.36).

(4.4) PROPOSITION. Let X be a measure space with a finite, positive measure μ and let Φ be a continuous $L_S(X,\mu)$ - representation of a topological vector space E. Then there exists a unique continuous linear operator $\Theta : E \to L_R^0(X,\mu)$ with $\Phi = e^{2\pi i\Theta}$.

Proof. For each fixed $u \in E$, the mapping $s \to \Phi(su)$ is a continuous $L_S(X,\mu)$ - representation of R. Thus, by (4.3), there is a unique function $\vartheta_u \in L_R^0(X,\mu)$ with

$$[\Phi(su)f](x) = f(x)\cdot\,\exp\,[2\pi is\vartheta_u(x)] \quad (s \in R;\ f \in L_C^2(X,\mu);\ x \in X).$$

From the uniqueness of ϑ in (4.3) it follows easily that $\vartheta_{su} = s\vartheta_u$ and $\vartheta_{u+v} = \vartheta_u + \vartheta_v$ for all $u,v \in E$ and $s \in R$. So, the mapping $\Theta : E \to L_R^0(X,\mu)$ given by $u \to \vartheta_u$ is linear. The continuity of Θ follows from (4.1). ▪

(4.5) PROPOSITION. Let K be a subgroup of a separable topological vector space E. If the quotient group E/K admits a non-trivial continuous unitary representation, then there exists a non-zero continuous linear operator $\Theta : E \to L_R^0(0,1)$ with $\Theta(K) \subset L_Z^0(0,1)$.

Proof. Let Φ be a non-trivial continuous unitary representation of E/K in a Hilbert space H. In view of (1.21), we may assume Φ to be cyclic. This implies that H is separable. Let A be the algebra of operators in H generated by operators Φ_g, $g \in E/K$. Then A is an abelian self-adjoint algebra in H containing the identity operator. The closure $\bar{A}$ of A in the strong operator topology is an abelian von Neumann algebra in H. Therefore we can decompose H into an at most countable Hilbert sum of $\bar{A}$-invariant subspaces H_n such that, for each n, either $\dim H_n = 1$ or the restriction of $\bar{A}$ to H_n is unitarily isomorphic to the algebra $L_C^\infty(0,1)$ in the Hilbert space

$L^2_C(0,1)$. The last sentence follows from the standard results on the structure of abelian von Neumann algebras; perhaps the best reference here will be [47], Section 9.4.

The subspaces H_n, being invariant for $\overline{A}$, are invariant for Φ. So, Φ can be decomposed into a Hilbert sum of some representations Φ_n which are either one-dimensional or unitarily equivalent to $L_S(0,1)$--representations. One of these representations is non-trivial, therefore we may assume that Φ itself is one-dimensional or an $L_S(0,1)$--representation. In the first case, E/K admits a non-trivial continuous character; let us denote it by χ. Let $\psi : E \to E/K$ be the natural projection. Then $\chi\psi \in E^{\wedge}$. By (2.3), we have $\chi\psi = \rho f$ for some $f \in E^*$. Since $\chi\psi \neq 0$ and $\chi\psi(K) = \{0\}$, it follows that $f \neq 0$ and $f(K) \subset Z$. Now, we may define Θ by the formula $(\Theta u)(x) = f(u)$ for $u \in E$ and $x \in (0,1)$.

It remains to consider the case when Φ is an $L_S(0,1)$ - representation. Then $\Phi\psi$ is an $L_S(0,1)$ - representation of E and, by (4.4), there exists a continuous linear operator $\Theta : E \to L^0_R(0,1)$ with $e^{2\pi i\Theta} = \Phi\psi$. It is clear that $\Theta \neq 0$ and $\Theta(K) \subset L^0_Z(0,1)$. ∎

A non-trivial Hausdorff abelian group is called <u>exotic</u> if it does not admit any non-trivial continuous unitary representations. We say that G is <u>strongly exotic</u> if it does not admit any weakly continuous representation in Hilbert spaces.

In connection with the above definition, let us notice that every topological group (abelian or not) admits a faithful strongly continuous representation by bounded operators in a suitably chosen Banach space. It suffices, for instance, to consider the representation by shift operators in the space of bounded and uniformly continuous functions on the group.

An abelian topological group G is said to be <u>bounded</u> if to each $U \in \mathcal{N}_0(G)$ there correspond a positive integer n and a finite subset A of G, such that $A + U^n = G$ (this definition makes sense also for non-abelian groups). For instance, all compact or connected groups are bounded. If K is a subgroup of a normed space E, the quotient group E/K is bounded if and only if there exists a number $r > 0$ such that $E = K + rB$ (cf. the proof of (5.1) (b) below).

(4.6) NOTE. In section 18, the expression "bounded group" will be used in a completely different meaning.

A representation Φ of a group G in a Banach space is called <u>bounded</u> if $\sup\{\|\Phi(g)\| : g \in G\} < \infty$.

(4.7) LEMMA. Every weakly continuous representation of a bounded and metrizable group in a Banach space is bounded.

Proof. Let Φ be a weakly continuous representation of a bounded and metrizable group G in a Banach space E. For simplicity, let us assume that G is abelian. Let (g_n) be an arbitrary null-sequence in G. For each $u \in E$, the sequence $(\Phi(g_n)u)_{n=1}^{\infty}$ converges weakly to u and is therefore bounded. Hence, by the Banach-Steinhaus theorem, we have

(1) $$\sup\{\|\Phi(g_n)\| : n = 1,2,\dots\} < \infty.$$

Since (g_n) was an arbitrary null-sequence and G is metrizable, from (1) it follows that there is some $U \in N_o(G)$ such that

$$C = \sup\{\|\Phi(g) : g \in U\} < \infty.$$

Since G is a bounded group, there are a positive integer n and a finite subset A of G, such that $A + U^n = G$. Then

$$\sup\{\|\Phi(g) : g \in G\} \leqq C^n \max\{\|\Phi(g)\| : g \in A\} < \infty. \quad \blacksquare$$

(4.8) LEMMA. Every bounded representation of an abelian group in a Hilbert space is equivalent to a unitary representation.

Proof. There is an invariant mean on the space of all bounded, real-valued functions on an abelian group ([38], (17.5)), and it is enough to repeat the standard argument for compact groups (cf. [53], Exercise 1 in sect. 9.3). $\blacksquare$

(4.9) PROPOSITION. Every non-trivial abelian group of automorphisms of a Banach space admits a non-trivial character continuous in the uniform topology.

Proof. Let G be a non-trivial abelian group of automorphisms of a Banach space and let A be the complex Banach algebra spanned over G. Let $\mathfrak{m}$ be the set of all multiplicative linear functionals of A. Suppose first that there are some $g \in G$ and $f \in \mathfrak{m}$ such that $f(g) \neq 1$. The multiplicative group $\mathbb{C} \setminus \{0\}$ admits a continuous character χ such that $\chi(f(g)) \neq 1$. Then χf is a non-trivial continuous character of G.

So, we may assume that $f(g) = 1$ for all $g \in G$ and $f \in \mathfrak{m}$. This implies that $G \subset e + \operatorname{rad} A$ where e is the unit of A and $\operatorname{rad} A$ the radical of A. Since $e + \operatorname{rad} A \subset \exp A$, we obtain $G \subset \exp A$. Thus we can find an element $a \in A$ such that $\exp a \in G$ and $\exp a \neq 1$. Then $a \notin 2\pi i Z e$ and the Hahn-Banach theorem implies the existence of a continuous R-linear functional $f : A \to R$ such that $f(2\pi i Z e) \subset Z$ and $f(a) \notin Z$.

Since span G is dense in A and all functionals in $\mathfrak{m}$ are trivial on G, it follows that $\mathfrak{m}$ consists of only one element. Therefore the exponential mapping is simply periodic, i.e. $\exp x = e$ implies that $x \in 2\pi i Z e$ (see e.g. [39], Sect. 5.6). Hence $\rho f(x) = 0$ if $\exp x = e$. Consequently, there exists a continuous homomorphism χ of the multiplicative group $\exp A$ into T such that the diagram

$$\begin{array}{ccc} A & \xrightarrow{\ f\ } & R \\ \Big\downarrow{\scriptstyle \exp} & & \Big\downarrow{\scriptstyle \rho} \\ \exp A & \xrightarrow{\ \chi\ } & T \end{array}$$

commutes. Moreover, $\chi(\exp a) = \rho(f(a)) \neq 0$. Thus $\chi_{|G}$ is a continuous non-trivial character of G. ∎

From (4.9) it follows that if an abelian topological group does not admit non-trivial continuous characters, then it does not admit any non-trivial uniformly continuous representations in Banach spaces.

(4.10) EXAMPLE. Let H be a complex Hilbert space with an orthonormal basis $(e_n)_{n=1}^{\infty}$. Let us set

$$e_1 e_n = e_n e_1 = e_n \quad \text{for} \quad n = 1,2,\dots;$$

$$e_2 e_n = e_n e_2 = 0 \quad \text{for} \quad n = 2,3,\dots;$$

$$e_n e_m = e_m e_n = 0 \quad \text{for} \quad m,n = 3,4,\dots;\ m \neq n;$$

$$e_n^2 = e_2 + e_n \quad \text{for} \quad n = 3,4,\dots .$$

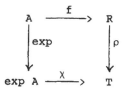

A direct verification shows that these formulae define on H a structure of a commutative Banach algebra with unit e_1. The multiplicative group $G = \exp H$ may be identified with $H/\exp^{-1}(e_1)$. A direct calculation shows that $\exp^{-1}(e_1)$ is generated by the elements $2\pi i e_1$ and $2\pi i(e_2 + e_n)$, $n = 3,4,\ldots$. So, $\exp^{-1}(e_1)$ is not weakly closed in H and (2.5) implies that continuous characters do not separate points of G (more precisely, $\chi(\exp 2\pi i e_2) = 0$ for all $\chi \in G^{\wedge}$).

Nevertheless, G admits a continuous, faithful unitary representation in a separable Hilbert space. The proof is similar to that of (5.1) (e) and consists in constructing a continuous linear operator $\Theta : H \to L^0_R(0,1)$ with the property that $\Theta u \in L^0_Z(0,1)$ if and only if $u \in \exp^{-1}(e_1)$.

(4.11) NOTE. The material of this section is standard, with the exception of (4.7), (4.9) and (4.10) which are taken from [9] and [6].

5. Quotients of normed spaces

By c_o and l^p, $1 \leq p < \infty$, we denote the classical Banach sequence spaces with their usual norms. By e_n, $n = 1,2,\ldots$, we denote the sequence $(0,\ldots,0,1,0,\ldots)$ with 1 in the nth place.

(5.1) THEOREM. Let $(a_n)_{n=1}^{\infty}$ be a sequence in l^1 such that

(1) $\quad a_n \in \operatorname{span}\{e_i\}_{i<n}$ for each $n = 1,2,\ldots$

(see (2.1)). Suppose that $\{a_n\}_{n=1}^{\infty}$ is a dense subset of l^1 and let $K = \operatorname{gp}\{a_n + e_n\}_{n=1}^{\infty}$. Then

(a) K is a discrete subgroup in each of the spaces c_o and l^p, $1 \leq p < \infty$;

(b) the groups c_o/K and l^p/K, $1 \leq p < \infty$, are bounded;

(c) the groups c_o/K and l^p/K, $p > 1$, do not admit any non--trivial continuous characters;

(d) the groups c_o/K and l^p/K, $p > 2$, are strongly exotic;

(e) the groups l^p/K, $1 \leq p \leq 2$, admit faithful continuous unitary representations in separable Hilbert spaces;

(f) l^1/K admits sufficiently many continuous characters.

Proof. Throughout the proof, E denotes one of the spaces c_o and l^p, $1 \leq p < \infty$, and $\psi : E \to E/K$ is the natural projection.

(a) Choose any $u \in K \setminus \{0\}$. We may write

$$u = k_1(a_1 + e_1) + \ldots + k_m(a_m + e_m)$$

for some $k_1,\ldots,k_m \in Z$ with $k_m \neq 0$. From (1) it follows that the mth coordinate of u is equal to k_m. Hence, according to the definition of the norm in E, we have $\|u\| \geq |k_m| \geq 1$.

(b) First, we shall prove that

$$(2) \qquad K + 2B_E = E.$$

Choose any $u \in E$. Since $\{a_n\}$ is dense in l^1, it is dense in E. So, there is some $n = 1,2,\ldots$ with $u \in a_n + B_E$. Then

$$u \in (a_n + e_n) - e_n + B_E \subset K + B_E + B_E = K + 2B_E,$$

which proves (2). Now, choose any $U \in N_o(E/K)$. We have $\psi(rB_E) \subset U$ for some $r > 0$. Take an integer m such that $mr \geq 2$. Then, by (2),

$$U^m \supset \psi(rB_E)^m = \psi((rB_E)^m) = \psi(mrB_E) \supset \psi(2B_E) = \psi(2B_E + K)$$
$$= \psi(E) = E/K.$$

(c) Let $E = c_o$ or $E = l^p$ with $p > 1$. Since $\{a_n\}$ is dense in E and (e_n) converges weakly to zero, it follows that K is weakly dense in E. Now, it remains to apply (2.5).

(d) Let $E = c_o$ or $E = l^p$ with $p > 2$. Suppose to the contrary that E/K is not strongly exotic. Then, in virtue of (b), (4.7), (4.8) and (4.5), there exists a non-zero continuous linear operator $\Theta : E \to L^0_R(0,1)$ with $\Theta(K) \subset L^0_Z(0,1)$. Since $\Theta \neq 0$, we can find some $u \in E$ such that the set

$$Q = \{t \in (0,1) : (\Theta u)(t) \in Z + [\tfrac{2}{5},\tfrac{3}{5}]\}$$

has a positive measure. Let (a_{k_n}) be a subsequence of (a_n) converging to u. The sequence (Θa_{k_n}) converges in measure to Θu. Applying the Egorov theorem, we may assume that $\Theta a_{k_n} \to \Theta u$ uniformly on Q. Hence

$$(\Theta a_{k_n})(t) \in Z + [\tfrac{1}{4},\tfrac{3}{4}]$$

for $t \in Q$ and n sufficiently large (say, for $n > n_o$). All functions $\Theta(a_{k_n} + e_{k_n})$ assume integer values only, therefore

$$(\Theta e_{k_n})(t) \in Z + [\tfrac{1}{4},\tfrac{3}{4}] \qquad \text{for } t \in Q \text{ and } n > n_o.$$

This implies that

$$(3) \qquad |\Theta e_{k_n})(t)| \geq \frac{1}{4} \qquad \text{for } t \in Q \text{ and } n > n_o.$$

Set $f_n = n^{-1/2}\Theta e_{k_n}$ for $n = 1,2,\dots$. Each subseries of the series $\sum_{n=1}^{\infty} n^{-1/2} e_{k_n}$ is convergent in E. So, by the continuity of Θ, each subseries of $\sum_{n=1}^{\infty} f_n$ is convergent in $L_R^0(0,1)$. Hence it follows that $\sum_{n=1}^{\infty} f_n^2(t) < \infty$ for almost all $t \in (0,1)$ (see [73], Lemma on p. 29). On the other hand, (3) implies that $\sum_{n=1}^{\infty} f_n^2(t) = \infty$ for $t \in Q$. The contradiction obtained completes the proof of (d).

(e) Let $E = l^p$ where $1 \leq p \leq 2$. Choose any $u = (x_n)_{n=1}^{\infty} \in E\backslash K$. Being discrete, K is closed in E. Moreover, we have

$$\sum_{n=1}^{m} x_n e_n \to u \quad \text{as} \quad m \to \infty.$$

Therefore we can find an index m such that

$$w := \sum_{n=1}^{m} x_n e_n \notin L := \text{gp}\,\{a_n + e_n\}_{n=1}^{m},$$

$$(4) \qquad \sum_{n=m+1}^{\infty} x_n^2 < \frac{1}{9}.$$

Let $M = \text{span}\,\{a_n + e_n\}_{n=1}^{m}$. Then L is a lattice in M and $w \notin L$, so that we can find some $f \notin M^*$ with $f(L) \subset Z$ and $f(w) \notin Z$. Multiplying f, if need be, by a suitable integer, we may assume that

$$(5) \qquad f(w) \in Z + [\tfrac{1}{3},\tfrac{2}{3}].$$

We shall construct a bounded linear operator $\Theta : E \to L_R^0(0,1)$ with $\Theta(K) \subset L_Z^2(0,1)$ and $\Theta u \notin L_Z^2(0,1)$. We may write

$$(6) \qquad a_n = \sum_{k=1}^{n-1} \alpha_{kn} e_k \qquad (n = 2,3,\dots)$$

for some coefficients α_{kn}. We shall construct step-functions $h_1, h_2, \ldots$ on $(0,1)$ such that $|h_{m+1}|, |h_{m+2}|, \ldots \leq 1$, the functions $h_{m+1}, h_{m+2}, \ldots$ are pairwise orthogonal and

$$(7) \qquad h_n + \sum_{k=1}^{n-1} \alpha_{kn} h_k \in L_Z^2(0,1)$$

for $n = 2,3,\ldots$. Set $h_n \equiv f(e_n)$ for $n = 1,\ldots,m$. The functions $h_{m+1}, h_{m+2}, \ldots$ will be constructed inductively.

Suppose we have constructed functions $h_1,\ldots,h_{n-1}$ for a certain $n \geq m+1$. The interval $(0,1)$ decomposes into a finite union of some intervals I_i such that, for each i, the functions $h_1,\ldots,h_{n-1}$ are constant on I_i. Fix an arbitrary index i and let p be the value of the function $\sum_{k=1}^{n-1} \alpha_{kn} h_k$ on I_i. Let us write $I_i = (a,b)$. If $p \in Z$, we set $h_n \equiv 0$ on I_i. If $p \notin Z$, then

$$c := b + (b-a)([p]-p) \in (a,b)$$

and we define

$$h_n(t) = \begin{cases} [p] - p & \text{for} \quad t \in (a,c) \\ [p] - p + 1 & \text{for} \quad t \in (c,b). \end{cases}$$

Here $[p]$ denotes the integer part of p. Then $|h_n(t)| \leq 1$ and

$$h_n(t) + \sum_{k=1}^{n-1} \alpha_{kn} h_k(t) = h_n(t) + p \in Z$$

for $t \in I_i$. Moreover, we have

$$\int_a^b h_n(t)dt = 0.$$

In the same way we define h_n on other intervals. Then (7) is satisfied and it is clear that h_n is orthogonal to $h_1,\ldots,h_{n-1}$.

The sequence $(h_n)_{n=1}^{\infty}$ has the desired properties. We only have to verify (7) for $n = 2,3,\ldots,m$. But, by (6), for any such n, we have

$$h_n + \sum_{k=1}^{n-1} \alpha_{kn} h_k \equiv f(e_n) + \sum_{k=1}^{n-1} \alpha_{kn} f(e_k) = f(e_n + \sum_{k=1}^{n-1} \alpha_{kn} e_k)$$

$$= f(e_n + a_n) \in Z.$$

Since $|h_n| \leq 1$ for $n > m$ and the functions $h_{m+1}, h_{m+2}, \ldots$ are pairwise orthogonal, it follows that the conditions $\Theta e_n = h_n$, $n = 1,2,\ldots$, define a bounded linear operator $\Theta : E \to L^2_R(0,1)$. From (7) we see that $\Theta(a_n + e_n) \in L^2_Z(0,1)$ for each $n = 1,2,\ldots$; consequently, $\Theta(K) \subset L^2_Z(0,1)$. On the other hand, from (5) we derive

$$\Theta w = \Theta \sum_{n=1}^{m} x_n e_n \equiv \sum_{n=1}^{m} x_n f(e_n) = f(w) \in Z + [\tfrac{1}{3},\tfrac{2}{3}]$$

and (4) yields

$$\|\Theta(u - w)\| = \|\Theta \sum_{n=m+1}^{\infty} x_n e_n\| = (\sum_{n=m+1}^{\infty} x_n^2)^{1/2} < \tfrac{1}{3},$$

which implies that Θu is not an integer-valued function.

We may treat Θ as a continuous linear operator from E into $L^0_R(0,1)$. Then, by (4.1), $e^{2\pi i\Theta}$ is a continuous unitary representation of E in the space $L^2_C(0,1)$. Since $e^{2\pi i\Theta}$ is trivial on K, the formula $\Phi\psi = e^{2\pi i\Theta}$ defines a continuous unitary representation Φ of E/K in $L^2_C(0,1)$, with $\Phi(\psi(g)) \neq 1$.

So, to each $g \in X := (E/K) \setminus \{0\}$ there corresponds a continuous unitary representation Φ_g of E/K in $L^2_C(0,1)$, with $\Phi_g(g) \neq 1$. Being continuous, Φ_g remains non-trivial on some neighbourhood U_g of g. But X is a Lindelöf space, therefore we can find a countable subset A of X such that $\{U_g\}_{g\in A}$ is a covering of X. Then $\bigoplus_{g\in A} \Phi_g$ is a continuous faithful unitary representation of E/K in a separable Hilbert space.

(f) In view of (2.5), it is enough to show that K is weakly closed in l^1. The proof is similar to that of (e) and we leave out the details, the more so that that we shall not use (f) in the sequel. ▪

(5.2) REMARK. A topological group is called <u>monothetic</u> if it contains a cyclic dense subgroup ([38], (9.2)). The groups c_o/K and l^p/K from (5.1) are monothetic. In fact, the following statement is true: if K is a linearly dense subgroup of a metrizable and complete vector space E, then E/K is a monothetic group. The proof, being not difficult, is omitted.

It follows from (5.1) (c) and (f) that the group K occurring there is weakly dense in l^p for $p > 1$, but weakly closed in l^1. This gives rise to the question whether each discrete (resp. closed) subgroup of l^1 is weakly closed (cf. [82], Problem 2). The answer is negative:

(5.3) THEOREM. Every infinite dimensional normed space E contains a free and discrete subgroup K such that E/K is strongly exotic.

The proof is given in [6]. We do not repeat it here because it is similar to that of (6.1), given below, and even a bit simpler.

(5.4) PROPOSITION. Let K be a free and discrete subgroup of a non-zero topological vector space E. If K is weakly dense in E, then K cannot be dually embedded.

Proof. Let $(e_i)_{i \in I}$ be a system of free generators of K. Let χ be the character of K given by $\chi(e_i) = \frac{1}{2}$ for $i \in I$. Suppose that χ can be extended to a continuous character $\tilde{\chi}$ of E. Then, due to (2.3), there is a continuous linear functional f on E with $\rho f = \tilde{\chi}$. For each $i \in I$, we have $\rho f(e_i) = \tilde{\chi}(e_i) = \chi(e_i) = \frac{1}{2}$, i.e.

$$(1) \qquad f(e_i) \in \frac{1}{2} + Z.$$

This means that $(2f)(K) \subset Z$. So, if K is weakly dense in E, then $2f = 0$, which contradicts (1). ∎

It follows from (5.4) that the group K from (5.3) is not dually embedded. Similarly, the group K from (5.1) is neither dually embedded in c_o nor in l^p, $p > 1$; it is clear, however, that K is dually embedded in l^1.

(5.5) REMARKS. The first example of a closed subgroup of a Banach space which is not weakly closed was given by Hooper [41], p. 254. Sidney [82] proved that if a Banach space has a separable infinite dimensional quotient space, then it contains a weakly dense proper closed subgroup. He also proved that if a Banach space X has a normalized basis $(e_n)_{n=1}^{\infty}$ such that e_n tends weakly to zero, then X contains a weakly dense discrete subgroup. His method is different from ours used in (5.1).

(5.6) REMARKS. The fact that the groups K in (5.1) and (5.3) are free is not accidental; Sidney [82], p. 983, proved that a countable and discrete subgroup of a normed space must be free. Countability is essential here: $L_Z^\infty(0,1)$ is a discrete but not free subgroup of $L_R^\infty(0,1)$. Conversely, a closed and free subgroup of a Banach space must be discrete. These facts were pointed out to the author by W. Wojtyński.

The example of an exotic group given by Herer and Christensen [37] was a separable, metrizable and complete topological vector space (the space of measurable functions relative to a certain pathological submeasure). Naturally, the space $L_R^0(0,1)$ is not an exotic group. We shall see, however, that it has an exotic quotient space.

We shall have to distinguish between a measurable function and its class of equivalence; the space of all real-valued measurable <u>functions</u> on $(0,1)$ will be denoted by $\mathcal{L}^0$ and the class of equivalence of a function $f \in \mathcal{L}^0$ by $\{f\}$. A sequence $(f_n)_{n=1}^\infty$ of functions belonging to $\mathcal{L}^0$ is said to be <u>pointwise linearly independent</u> if to each system $t_1,\dots,t_r \in (0,1)$ with $t_1 < \dots < t_r$ there correspond some indices $n_1,\dots,n_r$ such that $\det |f_{n_i}(t_j)|_{i,j=1}^r \neq 0$.

(5.7) **LEMMA.** Let $(f_n)_{n=1}^\infty$ be a pointwise linearly independent sequence in $\mathcal{L}^0$. If $\Theta : L_R^0(0,1) \to L_R^0(0,1)$ is a continuous linear operator with $\Theta\{f_n\} = 0$ for every n, then $\Theta = 0$.

Proof. Let us write $L^0 = L_R^0(0,1)$ and denote the Lebesgue measure on $(0,1)$ by λ. Let $(\phi_k)_{k=1}^\infty$ be a sequence of functions from $\mathcal{L}^0$. Suppose that the measure of the set

$$X = \{t : \phi_k(t) = 0 \quad \text{for almost all} \quad k\}$$

is equal to 1. Let then $(\Phi_k)_{k=1}^\infty$ be a sequence of measurable mappings from $(0,1)$ into itself, such that if A is a measurable subset of $(0,1)$ with $\lambda(A) = 0$, then $\lambda(\Phi_k^{-1}(A) \cap \{t : \phi_k(t) \neq 0\}) = 0$ for every k. Given a function $f \in \mathcal{L}^0$, we set

$$f_\Gamma(t) = \begin{cases} \sum_{k=1}^\infty \phi_k(t) f(\Phi_k(t)) & \text{if} \quad t \in X \\ 0 & \text{if} \quad t \notin X. \end{cases}$$

Then $f_\Gamma \in \mathcal{L}^0$ and the mapping $f \to f_\Gamma$ induces, as can easily be seen, a continuous linear operator $\Gamma : L^0 \to L^0$. It was proved by S. Kwapień

[56] that, conversely, each continuous linear operator from L^0 into itself can be represented in such a form. So, to prove our lemma, we have to show that $\Gamma = 0$ provided that $\Gamma\{f_n\} = 0$ for every n.

Let us suppose that the latter condition is satisfied. Then there is a subset Y of X with $\lambda(Y) = 1$, such that, for each $t \in Y$,

$$(1) \qquad \sum_{k=1}^{\infty} \phi_k(t) f_n(\Phi_k(t)) = 0 \qquad (n = 1,2,\ldots).$$

Let us fix an arbitrary $t \in Y$. Since $t \in X$, there exists an index m such that

$$(2) \qquad \sum_{k=1}^{\infty} \phi_k(t) f(\Phi_k(t)) = \sum_{k=1}^{m} \phi_k(t) f(\Phi_k(t))$$

for <u>any</u> function f on $(0,1)$. Hence, by (1),

$$(3) \qquad \sum_{k=1}^{m} \phi_k(t) f_n(\Phi_k(t)) = 0 \qquad (n = 1,2,\ldots).$$

Denote $r = \operatorname{card}\{\Phi_k(t)\}_{k=1}^m$ and choose indices $k_1,\ldots,k_r \in \{1,\ldots,m\}$ such that $\Phi_{k_1}(t) < \ldots < \Phi_{k_r}(t)$. Then we can find some coefficients $\xi_1,\ldots,\xi_r$ such that

$$(4) \qquad \sum_{k=1}^{m} \phi_k(t) f(\Phi_k(t)) = \sum_{j=1}^{r} \xi_j f(\Phi_{k_j}(t))$$

for any function f on $(0,1)$. Hence, by (3),

$$(5) \qquad \sum_{j=1}^{r} \xi_j f_n(\Phi_{k_j}(t)) = 0 \qquad (n = 1,2,\ldots).$$

Since the sequence (f_n) is pointwise linearly independent, we can find indices $n_1,\ldots,n_r$ such that $\det |f_{n_i}(\Phi_{k_j}(t))|_{i,j=1}^r \neq 0$. In view of (5), this implies that $\xi_1 = \ldots = \xi_r = 0$. Hence, by (2) and (4), we have $f_\Gamma(t) = 0$ for each $f \in L^0$. Since t was an arbitrary element of Y and $\lambda(Y) = 1$, it follows that $\Gamma\{f\} = 0$ for each $f \in L^0$, i.e. that $\Gamma = 0$. ■

In view of (5.7) and (4.5), to obtain an exotic quotient space of $L_R^0(0,1)$ it suffices to find a pointwise linearly independent sequence (f_n) of measurable functions such that their classes $\{f_n\}$ are not linearly dense in $L_R^0(0,1)$. There are many such sequences. Perhaps the simplest one is the sequence of functions $t \to t^{n_k}$ where $(n_k)_{k=1}^{\infty}$ is a sequence of positive numbers such that $n_{k+1}/n_k \geq 1 + a$ and a is a solution of the inequality $2a > (1 + a)^{1+1/a}$ $(a > 3.403\ldots)$. That the sequence (t^{n_k}) is pointwise linearly independent follows from the well-known fact that if $n_1 < \ldots < n_r$ and $t_1 < \ldots < t_r$, then $\det |t_j^{n_i}|_{i,j=1}^{r} > 0$ (see e.g. [77], Part 5, Problem 76). That the functions t^{n_k} are not linearly dense in $L_R^0(0,1)$ is a consequence of a result of S. Mazur [64].

(5.8) REMARK. Let $(n_k)_{k=1}^{\infty}$ be a sequence of positive integers such that $n_{k+1}/n_k \geq q$ for all k, where q is some fixed number larger than 1. From the theory of trigonometric series it follows that the closed linear subspace of $L_R^0(0,2\pi)$ spanned over all functions $\sin(n_k t)$ and $\cos(n_k t)$, $k = 1,2,\ldots$, consists of all functions of the form

$$\sum_{k=1}^{\infty} (a_k \cos(n_k t) + b_k \sin(n_k t))$$

where $\sum_{k=1}^{\infty} (a_k^2 + b_k^2) < \infty$, so that it is not the whole space $L_R^0(0,2\pi)$ (see, e.g. [105], Lemma (6.5), p. 203). It seems quite possible that, at least for certain sequences (n_k) with $n_{k+1}/n_k \geq q$, the system consisting of functions $\sin(n_k t)$ and $\cos(n_k t)$ is pointwise linearly independent on $(0,2\pi)$.

(5.9) NOTE. The material of this section is taken from [9]. Proposition (5.4), Lemma (5.7) and the example of an exotic quotient of L^0, subsequent to (5.7), are new.

6. Quotients of non-nuclear spaces

This section is devoted to the proof of the following fact:

(6.1) THEOREM. Let E be a metrizable locally convex space. If E is not nuclear, then it contains a discrete subgroup K such that the quotient group $(\operatorname{span} K)/K$ is exotic.

The proof will be preceded by several lemmas. We have to introduce some new notation. Let U,W be two convex bodies (compact convex sets with non-empty interiors) in an n-dimensional vector space N. Their volume ratio (relative to any translation-invariant measure on N) will be denoted by $\left|\frac{U}{W}\right|$. That is,

$$\left|\frac{U}{W}\right| = \frac{\operatorname{vol}\,\Omega(U)}{\operatorname{vol}\,\Omega(W)}$$

where vol is the Lebesgue measure on R^n and $\Omega : N \to R^n$ is any linear isomorphism.

Let $\Phi : E \to F$ be a bounded linear operator acting between normed spaces. For each $k = 1,2,\ldots,$ let us denote

$$v_k(\Phi : E \to F) = \sup_N \left[\frac{|\Phi(B_E \cap N)|}{|B_F \cap \Phi(N)|}\right]^{1/k}$$

where the supremum is taken over all linear subspaces N of E with $\dim N = \dim \Phi(N) = k$. If $\dim \Phi(E) < k$, we set $v_k(\Phi : E \to F) = 0$. If F is a subspace of some normed space F', then obviously, $v_k(\Phi : E \to F') = v_k(\Phi : E \to F)$ for every k. Therefore we may simply write $v_k(\Phi)$ instead of $v_k(\Phi : E \to F)$ (cf. the remarks before (2.9)). It is clear that $v_k(\Phi) \le \|\Phi\|$ for every k.

(6.2) NOTE. The numbers $v_k(\Phi)$ satisfy all conditions in Pietsch's definition of s-numbers ([76], 11.1.1) except monotonicity; it may happen that $v_{k+1}(\Phi) > v_k(\Phi)$ (see [59] and [3]).

(6.3) LEMMA. Let $\Phi_1,\ldots,\Phi_s$ be bounded operators for which the composition $\Phi_1 \ldots \Phi_s$ is defined. Then

$$v_k(\Phi_1 \ldots \Phi_s) \le v_k(\Phi_1) \ldots v_k(\Phi_s) \qquad (k = 1,2,\ldots).$$

Proof. It is enough to consider the case $s = 2$. So, suppose that we are given operators $\Phi_1 : E_2 \to E_1$ and $\Phi_2 : E_3 \to E_2$. Let N be an arbitrary subspace of E_3 with $\dim N = \dim (\Phi_1\Phi_2)(N) = k$. Then $\dim \Phi_2(N) = \dim \Phi_1(\Phi_2(N)) = k$ and we may write

$$\left[\frac{|(\Phi_1\Phi_2(B(E_3) \cap N)|}{|B(E_1 \cap (\Phi_1\Phi_2)(N)|}\right]^{1/k}$$

$$= \left[\frac{|\Phi_1(\Phi_2(B(E_3) \cap N))|}{|\Phi_1(B(E_2) \cap \Phi_2(N))|} \cdot \frac{|\Phi_1(B(E_2) \cap \Phi_2(N))|}{|B(E_1) \cap \Phi_1(\Phi_2(N))|}\right]^{1/k}$$

$$= \left[\frac{|\Phi_2(B(E_3) \cap N)|}{|B(E_2) \cap \Phi_2(N)|}\right]^{1/k} \cdot \left[\frac{|\Phi_1(B(E_2) \cap \Phi_2(N))|}{|B(E_1) \cap \Phi_1(\Phi_2(N))|}\right]^{1/k}$$

$$\le v_k(\Phi_2)\cdot v_k(\Phi_1).$$

Since N was arbitrary, it follows that $v_k(\Phi_1\Phi_2) \le v_k(\Phi_1)\cdot v_k(\Phi_2)$. ∎

Let $\Phi : E \to F$ be a bounded operator acting between normed spaces. By $h_k(\Phi : E \to F)$, $k = 1,2,\ldots,$ we denote the Hilbert numbers of Φ (see [76], 11.4).

(6.4) LEMMA. For every bounded operator $\Phi : E \to F$, one has

$$v_k(\Phi : E \to F) \ge h_k(\Phi : E \to F) \qquad (k = 1,2,\ldots).$$

Proof. Take any operators χ,Ψ with $\|\chi\|,\|\Psi\| \le 1$ such that the composition $\chi\Phi\Psi$ is defined and acts between unitary spaces. Fix an arbitrary $k = 1,2,\ldots$. We have to show that $v_k(\Phi) \ge d_k(\chi\Phi\Psi)$. Denote $\Theta = \chi\Phi\Psi$. Since Θ acts between unitary spaces, we have $v_k(\Theta) = [d_1(\Theta) \ldots d_k(\Theta)]^{1/k}$. Hence, applying (6.3) with $s = 3$, we derive

$$d_k(\Theta) \le [d_1(\Theta) \ldots d_k(\Theta)]^{1/k} = v_k(\chi\Phi\Psi) \le v_k(\chi)v_k(\Phi)v_k(\Psi)$$

$$\le \|\chi\|\|\Psi\|v_k(\Phi) \le v_k(\Phi). \quad ∎$$

(6.5) LEMMA. Let E be a locally convex space. Suppose that there exists an $\varepsilon > 0$ such that to each continuous seminorm p on E there corresponds another seminorm $p' \ge p$ with $v_k(\Lambda_{p'p}) = o(k^{-\varepsilon})$. Then E is a nuclear space.

Proof. Choose an arbitrary continuous seminorm p_o on E. Next, take an integer $s > 5\varepsilon^{-1}$. Due to our assumptions, we can find continuous seminorms $p_s \ge \ldots \ge p_1 \ge p_o$ such that

$$v_k(\Lambda_{p_i p_{i-1}})) = o(k^{-\varepsilon}) \qquad (i = 1,\ldots,s).$$

Hence, by (6.3),

$$(1)\qquad v_k(\Lambda_{p_s p_o}) \le v_k(\Lambda_{p_s p_{s-1}}) \ldots v_k(\Lambda_{p_1 p_o}) = o(k^{-\varepsilon s}).$$

Let us denote $\Lambda = \Lambda_{p_s p_o} : E_{p_s} \to E_{p_o}$. From the inequality

$$\prod_{j=1}^{k} d_j(\Lambda) \le e^k k! \prod_{j=1}^{k} h_j(\Lambda)$$

([76], 11.12.3) and from (6.4) we derive

$$d_k(\Lambda) \le [\prod_{j=1}^{k} d_j(\Lambda)]^{1/k} \le e^k k! [\prod_{j=1}^{k} h_j(\Lambda)]^{1/k} \le ek[\prod_{j=1}^{k} v_j(\Lambda)]^{1/k}$$
$$(k = 1,2,\ldots).$$

In view of (1) and the inequality $\varepsilon s > 5$, this implies that $d_k(\Lambda) = o(k^{-4})$. Consequently, Λ is a nuclear operator ([79], Proposition 7.2.2). We have thus shown that, for each continuous seminorm on E, there exists another seminorm such that the corresponding operator is nuclear. This means that E is a nuclear space ([79], Theorem 7.2.7). ▪

(6.6) LEMMA. Let $U \subset W$ be two symmetric convex bodies in some n--dimensional vector space N. Let M be some m-dimensional subspace of N. Then

$$\frac{|U \cap M|}{|W \cap M|} \ge \frac{m!}{n!} \cdot \frac{|U|}{|W|}.$$

Proof. Suppose first that $m = n - 1$. Then we may assume that $N = R^n$ and

$$M = R^{n-1} := \{(x_1,\ldots,x_n) \in R^n : x_n = 0\}.$$

Set

$$h(U) = \sup\{x_n : (x_1,\ldots,x_n) \in U\},$$

$$h(W) = \sup\{x_n : (x_1,\ldots,x_n) \in W\}.$$

It is clear that

$$\mathrm{vol}_n(U) \le h(U)\,\mathrm{vol}_{n-1}(U \cap R^{n-1}),$$

$$\mathrm{vol}_n(W) \ge \tfrac{1}{n} h(W)\,\mathrm{vol}_{n-1}(W \cap R^{n-1}).$$

Hence

$$\frac{|U \cap M|}{|W \cap M|} = \frac{\text{vol }(U \cap M)}{\text{vol }(W \cap M)} \geq \frac{h(W)\ \text{vol }(U)}{nh(U)\ \text{vol }(W)} \geq \frac{1}{n}\frac{\text{vol }(U)}{\text{vol }(W)} = \frac{(n-1)!}{n!}\ \frac{|U|}{|W|}.$$

Now, suppose that $m < n - 1$. Then we can find linear subspaces $M = M_m \subset M_{m+1} \subset \ldots \subset M_n = N$ with $\dim M_{i+1} = 1 + \dim M_i$ for $i = m, m+1,\ldots,n$ and, due to the above, we have

$$\begin{aligned}\frac{|U \cap M|}{|W \cap M|} &\geq \frac{m!}{(m+1)!} \cdot \frac{|U \cap M_{m+1}|}{|W \cap M_{m+1}|} \\ &\geq \frac{m!}{(m+1)!} \cdot \frac{(m+1)!}{(m+2)!} \cdot \frac{|U \cap M_{m+2}|}{|W \cap M_{m+2}|} \geq \ldots \\ &\geq \frac{m!}{(m+1)!} \cdot \frac{(m+1)!}{(m+2)!} \cdots \frac{(n-1)!}{n!}\ \frac{|U \cap M_n|}{|W \cap M_n|} \\ &= \frac{m!}{n!}\ \frac{|U \cap N|}{|W \cap N|}. \quad \blacksquare\end{aligned}$$

(6.7) LEMMA. Let M be a finite dimensional subspace of an infinite dimensional normed space E. Then there exists a subspace N of E with $\operatorname{codim} N < \infty$, such that

$$\|u + v\| \geq \tfrac{1}{2}\|v\| \qquad \text{for all } u \in M \text{ and } v \in N.$$

The proof is standard.

(6.8) LEMMA. Let E,F be normed spaces and let $\Phi : E \to F$ be an injective bounded linear operator such that

(1) $$\limsup_{k\to\infty} k^{1/2} v_k(\Phi) = \infty.$$

If N is a subspace of E with $\operatorname{codim} N < \infty$, then

(2) $$\limsup_{k\to\infty} k^{1/2} v_k(\Phi_{|N}) = \infty.$$

Proof. Denote $s = \operatorname{codim} N$. If L is a k-dimensional subspace of E, then $l := \dim (L \cap N) \geq k - s$ and (6.6) yields

$$A := l^{1/2} \left[\frac{|\Phi(B_E \cap L \cap N)|}{|B_F \cap \Phi(L \cap N)|}\right]^{1/l} = l^{1/2}\left[\frac{|\Phi(B_E \cap L) \cap N|}{|B_F \cap \Phi(L) \cap N|}\right]^{1/l}$$

$$\geq \left(\frac{1}{k}\right)^{1/2} k^{1/2} \left[\frac{1!}{k!} \frac{|\Phi(B_E \cap L)|}{|B_F \cap \Phi(L)|}\right]^{1/1} .$$

Let us denote

$$Q = \frac{|\Phi(B_E \cap L)|}{|B_F \cap \Phi(L)}.$$

Then we may write

$$A \geq \left(\frac{k - s}{k}\right)^{1/2} k^{1/2} \left[\frac{(k - s)!}{k!}\right]^{1/(k-s)} Q^{1/(k-s)}$$

$$= \left(\frac{k - s}{k}\right)^{1/2} \left[\frac{(k - s)!}{k!}\right]^{1/(k-s)} k^{-s/2(k-s)} [k^{1/2} Q^{1/k}]^{k/(k-s)} .$$

Thus, if k and $k^{1/2} Q^{1/k}$ are both large, so is A (here s is fixed). In view of (1), this implies (2). ∎

(6.9) LEMMA. Let U be a symmetric convex body in R^n. If f is a linear functional on R^n such that

$$\mathrm{vol}_n (\{u \in U : |f(u)| \leq 1\}) \geq \frac{1}{2} \mathrm{vol}_n (U),$$

then

$$\mathrm{vol}_n (U)^{-1} \int_U f(u)^2 du < 7.$$

This is an easy consequence of the Brunn-Minkowski inequality (see [31], Corollary to Statement 2.1).

(6.10) LEMMA. Let U be a symmetric convex body in R^n. Let L be a lattice in R^n such that

(1) $\quad d(L) < \omega_n^{-2}(n + 2)^{-n/2} \mathrm{vol}_n (U)$

and let $a_1,\ldots,a_n$ be some fixed basis of L. Then there exist real numbers $c_1,\ldots,c_n$ such that if f is a linear functional on R^n with

(2) $\quad f(a_i) \in c_i + Z \quad$ for all $i = 1,\ldots,n,$

then

(3) $\quad \mathrm{vol}_n (\{u \in U : |f(u)| \geq \frac{1}{3}\}) \geq \frac{1}{2} \mathrm{vol}_n (U).$

Proof. Let $\|\ \|$ be the norm on R^n defined by

$$\|v\|^2 = \frac{1}{\mathrm{vol}_n(U)} \int_U (v,u)^2 du, \qquad v \in R^n,$$

and let D be the corresponding unit ball. Then D is an ellipsoid and $\mathrm{vol}_n(U)^{-1/2}D$ is the so-called Binet ellipsoid of U (see [65]). Let C be the Legendre ellipsoid of U, defined by the condition

$$\int_U (v,u)^2 du = \int_C (v,u)^2 du \qquad \text{for all } v \in R^n.$$

The connection between Legendre and Binet ellipsoids as well known in mechanics (see [42]); one has

$$(4) \qquad \mathrm{vol}_n(U)^{-1/2}D = \left(\frac{n+2}{\mathrm{vol}_n(C)}\right)^{1/2} C^0$$

where

$$C^0 = \{u \in R^n : (u,w) \le 1 \text{ for all } w \in C\}.$$

Blaschke [22] proved that $\mathrm{vol}_n(C) \ge \mathrm{vol}_n(U)$ (see also [42]). From this and (4) we get

$$\begin{aligned}
\mathrm{vol}_n(D) &= (n+2)^{n/2} \left(\frac{\mathrm{vol}_n(U)}{\mathrm{vol}_n(C)}\right)^{n/2} \mathrm{vol}_n(C^0) \\
&\le (n+2)^{n/2} \mathrm{vol}_n(C^0) \\
&= (n+2)^{n/2} \mathrm{vol}_n(C)\, \mathrm{vol}_n(C^0) \frac{\mathrm{vol}_n(U)}{\mathrm{vol}_n(C)} \frac{1}{\mathrm{vol}_n(U)} \\
&\le (n+2)^{n/2} \omega_n^2 \frac{1}{\mathrm{vol}_n(U)}.
\end{aligned}$$

Hence, by (1),

$$\mathrm{vol}_n(D) < d(L)^{-1} = d(L^*).$$

This implies that $L^* + D \ne R^n$ (see (3.21)). Choose some $w \notin L^* + D$ and set $c_i = (w, a_i)$ for $i = 1,\dots,n$.

Now, let f be a linear functional on R^n satisfying (2). Let v be the vector defined by $f(u) = (v,u)$ for all $u \in R^n$. Then, by (2),

$$(v - w, a_i) = (v, a_i) - (w, a_i) = f(a_i) - c_i \in Z$$

for $i = 1,\ldots,n$, which means that $v - w \in L^*$. Consequently, $v \notin D$, so that

$$\frac{1}{\mathrm{vol}_n\,(U)} \int_U f^2(u)du = \frac{1}{\mathrm{vol}_n\,(U)} \int_U (v,u)^2 du = \|v\|^2 > 1.$$

Inequality (3) follows now from (6.9). ∎

Proof of (6.1). Let $p_o \le p_1 \le \ldots$ be a sequence of seminorms defining the topology of E. More precisely, we assume that $\{B(p_n)\}_{n=0}^{\infty}$ is a base at zero in E. Due to (6.5), we may assume that

$$(1) \qquad \limsup_{k\to\infty} k^{1/2} v_k(\Lambda_{p_n p_o} : E_{p_n} \to E_{p_o}) = \infty$$

for $n = 1,2,\ldots$. Suppose first that all p_n´s are norms. We shall construct inductively a sequence $M_o, M_1, M_2, \ldots$ of finite dimensional subspaces of E.

Set $M_o = \{0\}$. Next, suppose that $M_o, M_1, \ldots, M_{n-1}$ have been constructed. Due to (6.7), there is a subspace N_n of E with $\operatorname{codim} N_n < \infty$, such that

$$(2) \qquad p_o(x + y) \ge \tfrac{1}{2} p_o(y) \quad \text{for all} \quad x \in M + \ldots + M_{n-1} \quad \text{and} \quad y \in N_n.$$

From (1) and (6.8) it follows that

$$\limsup_{k\to\infty} k^{1/2} v_k(\Lambda_{p_n p_o} | N_n) = \infty.$$

So, we can find a finite dimensional subspace M_n of N_n such that

$$(3) \qquad k(n) := \dim M_n \ge n,$$

$$(4) \qquad k(n)^{1/2} \left[\frac{|M_n \cap B(p_n)|}{|M_n \cap B(p_o)|}\right]^{1/k(n)} > 2\pi e,$$

and we may continue the induction.

After constructing the subspaces M_n we proceed to the construction of generators of the subgroup K. Let us fix an arbitrary $n = 1,2,\ldots$. We may identify M_n with $R^{k(n)}$; then $U_o := M_n \cap B(p_o)$ and $U_n := M_n \cap B(p_n)$ are bounded, symmetric and convex subsets of $R^{k(n)}$, and (4) says that

$$(5) \qquad \frac{\mathrm{vol}_{k(n)}\,(U_n)}{\mathrm{vol}_{k(n)}\,(U_o)} > (2\pi e)^{k(n)} k(n)^{-k(n)/2}.$$

According to the Minkowski-Hlawka theorem ([33], p. 202 or [28], Ch. VI, §3), we can find a lattice L_n in $M_n = R^{k(n)}$ such that

(6) $L_n \cap U_o = \{0\}$,

(7) $d(L_n) < \mathrm{vol}_{k(n)}(U_o)$.

Let $(a_{ni})_{i=1}^{k(n)}$ be some fixed basis of L_n. After easy calculations, from (7) and (5) we obtain

$$d(L_n) < \omega_{k(n)}^{-2}[k(n) + 2]^{-k(n)/2}\,\mathrm{vol}_{k(n)}(U_n).$$

So, by virtue of (6.10), there exist real coefficients c_{ni} with $i = 1,\dots,k(n)$, such that

(*) if f is a linear functional on M_n with $f(a_{ni}) \in c_{ni} + Z$ for $i = 1,\dots,k(n)$, then

(8) $\mathrm{vol}_{k(n)}(\{u \in U_n : |f(u)| \geq \frac{1}{3}\}) \geq \frac{1}{2}\,\mathrm{vol}_{k(n)}(U_n)$.

Now, an easy inductive argument allows us to construct vectors

(9) $w_{ni} \in M_o + \dots + M_{n-1}$ $(i = 1,\dots,k(n);\ n = 1,2,\dots)$

satisfying the following condition:

(**) given arbitrary integers $n_o \geq 0$, $p \neq 0$, $m \geq 1$ and $j = 1,\dots,k(n)$, there is an index $n > n_o$ such that

(10) $w_{ni} = -\frac{1}{p}c_{ni}(w_{mj} + a_{mj})$ $(i = 1,\dots,k(n))$.

Let K be the subgroup of E generated by all vectors $w_{ni} + a_{ni}$ where $n = 1,2,\dots$ and $i = 1,\dots,k(n)$. We shall prove that K is discrete. Choose any $u \in K \setminus \{0\}$. For a certain $n \geq 1$, we may write

$$u = v + \sum_{i=1}^{k(n)} r_i(w_{ni} + a_{ni})$$

where $v \in M_o + \dots + M_{n-1}$, and r_i are integers not all equal to zero. By (2) and (9), we have

$$p_o(u) = p_o(v + \sum_{i=1}^{k(n)} r_i w_{ni} + \sum_{i=1}^{k(n)} r_i a_{ni}) \geq \frac{1}{2}p_o(\sum_{i=1}^{k(n)} r_i a_{ni}).$$

From (6) we get $\sum_{i=1}^{k(n)} r_i a_{ni} \notin B(p_o)$, which implies that $p_u(u) > \frac{1}{3}$.

Finally, denote $F = \text{span } K$ and suppose that F/K admits a non--trivial continuous unitary representation. Then, due to (4.5), there exists a continuous non-zero linear operator $\Theta : F \to L_R^0(0,1)$ with $\Theta(K) \subset L_Z^0(0,1)$. For each $n = 1,2,\ldots,$ we have

(11) $$\Theta(w_{ni} + a_{ni}) \in L_Z^0(0,1) \qquad (i = 1,\ldots,k(n)).$$

For each pair n,i with $n = 1,2,\ldots$ and $i = 1,\ldots,k(n)$, choose a measurable function ϕ_{ni} on $(0,1)$, the class of which is equal to $\Theta(w_{ni} + a_{ni})$. By (11), we may assume that

(12) $$\phi_{ni}(t) \in Z \qquad (i = 1,\ldots,k(n); \ \ t \in (0,1))$$

for $n = 1,2,\ldots$. It follows from our construction that the vectors $w_{ni} + a_{ni}$ form a Hamel basis in F. Consequently, each vector $u \in F$ can be written in the form

(13) $$u = \sum_{n=1}^{\infty} \sum_{i=1}^{k(n)} x_{ni}(w_{ni} + a_{ni})$$

with all but finitely many coefficients x_i equal to zero. Naturally, such a representation is unique. For $t \in (0,1)$, let us write

$$f_t(u) = \sum_{n=1}^{\infty} \sum_{i=1}^{k(n)} x_{ni}\phi_{ni}(t).$$

It is clear that f_t is a linear functional on F. If u is given by (13), then

$$\Theta(u) = \sum_{n=1}^{\infty} \sum_{i=1}^{k(n)} x_{ni}\Theta(w_{ni} + a_{ni}).$$

This means that, for each fixed $u \in F$, the class of the function $t \to f_t(u)$ is equal to $\Theta(u)$. Observe that, for every n,

(14) $$f_t(w_{ni} + a_{ni}) = \phi_{ni}(t) \qquad (i = 1,\ldots,k(n); \ \ t \in (0,1)).$$

Since $\Theta \neq 0$, and $\{w_{ni} + a_{ni}\}$ is a Hamel basis in F, there are some $m = 1,2,\ldots$ and $j = 1,\ldots,k(m)$ with $\Theta(w_{mj} + a_{mj}) \neq 0$. Let λ be the Lebesgue measure on $(0,1)$. Replacing n by m in (12), we see that there exist a measurable subset X of $(0,1)$ with $\lambda(X) > 0$ and an integer $p \neq 0$, such that

(15) $\phi_{mj}(t) = p$ for all $t \in X$.

Take an arbitrary integer $n_o > 0$. By virtue of (**), there is an index $n > n_o$ such that (10) is satisfied. Let us write $k = k(n)$. Next, fix an arbitrary $t \in X$. From (10) we have

$$f_t(w_{ni}) = - \frac{1}{p} c_{ni} f_t(w_{mj} + a_{mj}) \qquad (i = 1,\dots,k).$$

Hence, replacing n by m in (14) and using (15), we get

$$(16) \qquad f_t(w_{ni}) = - \frac{1}{p} c_{ni} \phi_{mj}(t) = -c_{ni} \qquad (i = 1,\dots,k).$$

From (14) and (12) we derive

$$f_t(w_{ni} + a_{ni}) = \phi_{ni}(t) \in Z \qquad (i = 1,\dots,k).$$

Thus, by (16),

$$(17) \qquad f_t(a_{ni}) \in - f_t(w_{ni}) + Z \subset c_{ni} + Z \qquad (i = 1,\dots,k).$$

Let us treat f_t as a linear functional on M_n. From (*) and (17) it follows that

$$\mathrm{vol}_k\ (\{u \in U_n : |f_t(u)| \geq \tfrac{1}{3}\}) \geq \tfrac{1}{2}\ \mathrm{vol}_k\ (U_n).$$

Since this holds for each $t \in X$, from the Fubini theorem we infer that there exists a vector $u_n \in U_n$ such that

$$\lambda(\{t \in (0,1) : |f_t(u_n)| \geq \tfrac{1}{3}\}) \geq \tfrac{1}{2}\lambda(X).$$

Consequently, we have

$$(18) \qquad |\Theta(u_n)| = \int_X \min\,(1,|f_t(u_n)|)dt \geq \tfrac{1}{6}\lambda(X).$$

We have thus shown that to each $n_o = 1,2,\dots$ there corresponds some vector $u_n \in B(p_n) \subset B(p_{n_o})$ such that (18) is satisfied. Since $\{F \cap B(p_n)\}_{n=1}^{\infty}$ is a base at zero in F, it follows that Θ cannot be continuous. The contradiction obtained completes the proof.

If p_n's are not norms, the above argument requires only small technical modifications. ▪

(6.11) REMARK. The assumption of metrizability in (6.1) is essential. Let E be an infinite dimensional vector space with a countable Hamel basis. The topology of E is defined by some norm and by the family of all seminorms $u \mapsto |f(u)|$ where $f \in E^{\#}$. Naturally, E is a non-nuclear locally convex space. An easy argument shows that each closed subgroup of E is dually closed and dually embedded (see [4], p. 132).

(6.12) NOTE. Theorem (6.1) is new. The main idea of the proof, as well as Lemmas (6.3) - (6.7), are taken from [4].

Chapter 3

NUCLEAR GROUPS

7. Nuclear groups

(7.1) DEFINITION. A Hausdorff abelian group G is called nuclear if it satisfies the following condition:

given an arbitrary $U \in \mathcal{N}_o(G)$, $c > 0$ and $m = 1,2,\ldots,$ there exist: a vector space E, two symmetric and convex subsets X,Y of E with

$$(1) \qquad d_k(X,Y) \leq ck^{-m} \qquad (k = 1,2,\ldots),$$

a subgroup K of E and a homomorphism $\phi : K \to G$, such that $\phi(K \cap X) \in \mathcal{N}_o(G)$ and $\phi(K \cap Y) \subset U$.

In other words, G is nuclear if each neighbourhood of zero contains another neighbourhood which is "sufficiently small" with respect to the original one. From the proof of (9.6) it follows that (1) may be replaced by the condition

$$d_k(X,Y) \leq c_o k^{-m_o} \qquad (k = 1,2,\ldots)$$

where c_o and m_o are some universal constants. One may take, for instance, $c_o = 10^{-2}$ and $m_o = 4$. The author does not know whether it is sufficient to take, say, $c_o = 1$ and $m_o = 1$ (cf. (9.2) and (9.3)).

The above definition is rather complicated. However, it appears in a very natural way when one tries to define "intrinsically" a reasonable class of abelian topological groups which would include subgroups and Hausdorff quotients of nuclear spaces. For a simpler definition, see (9.9).

(7.2) PROPOSITION. A Hausdorff abelian group G is nuclear if and only if it satisfies the following condition:

given arbitrary $U \in \mathcal{N}_o(G)$, $c > 0$ and $m = 1,2,\ldots,$ there exist: a vector space E, two pre-Hilbert seminorms p,q on E with

$$d_k(B_p,B_q) \leq ck^{-m} \qquad (k = 1,2,\ldots),$$

a subgroup K of E and a homomorphism $\phi : K \to G$, such that $\phi(K \cap B_p) \in N_o(G)$ and $\phi(K \cap B_q) \subset U$.

Proof. The "if" part is trivial. The "only if " part follows directly from (2.14). ∎

(7.3) **PROPOSITION.** Discrete abelian groups are nuclear.

Proof. Set $E = \{0\}$ in (7.1). ∎

(7.4) **PROPOSITION.** Every nuclear locally convex space is a nuclear group.

Proof. Let E be a nuclear space. Choose arbitrary $c > 0$, $m = 1,2,\ldots$ and $U \in N_o(E)$. By (2.17), we can find symmetric and convex sets $W,V \in N_o(E)$ such that $W \subset V \subset U$ and

$$d_k(W,V) \leq ck^{-m} \qquad (k = 1,2,\ldots).$$

Now, we may take $X = W$, $Y = V$, $K = E$ and $\phi = id_E$ in (7.1). ∎

(7.5) **PROPOSITION.** Subgroups and Hausdorff quotient groups of nuclear groups are nuclear.

Proof. Let H be a subgroup of a nuclear group G. Choose arbitrary $c > 0$, $m = 1,2,\ldots$ and $U \in N_o(G)$. Next, take E,X,Y,K and ϕ as in (7.1).

Set $K' = \phi^{-1}(H)$ and $\phi' = \phi_{|K'}$. Then $\phi'(K \cap Y) \subset U \cap H$ and $\phi'(K' \cap X) = H \cap \phi(K \cap X) \in N_o(H)$. Since $\{U \cap H : U \in N_o(G)\} = N_o(H)$, it follows that H is a nuclear group.

Suppose now that H is closed and let $\psi : G \to G/H$ be the natural projection. Denote $\phi'' = \psi\phi$. Then $\phi''(K \cap Y) \subset \psi(U)$ and $\phi''(K \cap X) \in N_o(G/H)$ because ψ is open. Since $\{\psi(U) : U \in N_o(G)\} = N_o(G/H)$, it follows that G/H is a nuclear group. ∎

(7.6) **PROPOSITION.** The product of an arbitrary family of nuclear groups is a nuclear group.

Proof. Let G be the product of a family $\{G_i\}_{i \in I}$ of nuclear groups. Take arbitrary $c > 0$, $m = 1,2,\ldots$ and $U \in N_o(G)$. We can find a finite subset J of I and, for each $i \in J$, some $U_i \in N_o(G_i)$

such that $\prod_{i\in J} U_i \times \prod_{i\notin J} G_i \subset U$. We may write $J = \{i_n\}_{n=1}^{p}$. By (7.1), for each $n = 1,\dots,p$, we can find a vector space E_n, two symmetric and convex subsets X_n, Y_n of E_n with

$$d_k(X_n,Y_n) < c2^{-mn}k^{-m} \qquad (k = 1,2,\dots),$$

a subgroup K_n of E_n and a homomorphism $\phi_n : K_n \to G_{i_n}$ such that $\phi_n(K_n \cap X_n) \in N_o(G_{i_n})$ and $\phi_n(K_n \cap Y_n) \in U_{i_n}$. For each $i \in I \setminus J$, there exist a vector space E_i, a subgroup K_i of E_i and a surjective homomorphism $\phi_i : K_i \to G_i$ (every group is a quotient of a free one; a free group is a direct sum of Z's which, in turn, is a subgroup of the corresponding direct sum of R's). Set

$$E = \prod_{n=1}^{p} E_n \times \prod_{i\notin J} E_i, \quad X = \prod_{n=1}^{p} X_n \times \prod_{i\notin J} E_i, \quad Y = \prod_{n=1}^{p} Y_n \times \prod_{i\notin J} E_i.$$

Then X,Y are symmetric and convex subsets of the vector space E. From (2.7) we get

$$d_k(X,Y) \le ck^{-m} \qquad (k = 1,2,\dots).$$

Next, define

$$K = \prod_{n=1}^{p} K_n \times \prod_{i\notin J} K_i \qquad \text{and} \qquad \phi = (\phi_n)_{n=1}^{p} \times (\phi_i)_{i\notin J} : K \to G.$$

It is clear that $\phi(K \cap Y) \subset U$. On the other hand, we have

$$\prod_{n=1}^{p} \phi_n(K_n \cap X_n) \times \prod_{i\notin J} G_i \subset \phi(K \cap X),$$

whence $\phi(K \cap X) \in N_o(G)$. ∎

(7.7) **PROPOSITION.** The limit of an inverse system of nuclear groups is a nuclear group.

This is a direct consequence of (7.5) and (7.6).

(7.8) **PROPOSITION.** The direct sum of a countable family of nuclear groups is a nuclear group.

Proof. Let $(G_n)_{n=1}^{\infty}$ be a sequence of nuclear groups and let $G = \sum_{n=1}^{\infty} G_n$. Take arbitrary $c > 0$, $m = 1,2,\ldots$ and $U \in N_o(G)$. There are some $U_n \in N_o(G_n)$, $n = 1,2,\ldots$, with $\sum_{n=1}^{\infty} U_n \subset U$. By (7.1), for every n, we can find a vector space E_n, two symmetric and convex subsets X_n, Y_n of E_n with

$$d_k(X_n,Y_n) < c2^{-mn}k^{-m} \qquad (k = 1,2,\ldots),$$

a subgroup K_n of E_n and a homomorphism $\phi_n : K_n \to G_n$, such that $\phi_n(K_n \cap X_n) \in N_o(G_n)$ and $\phi_n(K_n \cap Y_n) \subset U_n$. Let us define

$$E = \sum_{n=1}^{\infty} E_n, \qquad X = \sum_{n=1}^{\infty} X_n, \qquad Y = \sum_{n=1}^{\infty} Y_n.$$

From (2.7) we get $d_k(X,Y) \leq ck^{-m}$ for every k. Setting $K = \sum_{n=1}^{\infty} K_n$ and $\phi = (\phi_n)_{n=1}^{\infty} : K \to G$, we have $\phi(K \cap Y) \subset U$ and $\sum_{n=1}^{\infty} \phi_n(K_n \cap X_n) \subset \phi(K \cap X)$, whence $\phi(K \cap X) \in N_o(G)$. ∎

(7.9) PROPOSITION. The limit of a countable direct system of nuclear groups is a nuclear group provided that it is separated.

This follows directly from (7.5) and (7.8).

The limit of a direct system of nuclear groups (and even of LCA groups) need not be separated. See, however, (1.18) and the remarks following the definition of the direct system of groups in section 1.

From (7.5) - (7.9) we see that the permanence properties of nuclear groups are similar to those of nuclear spaces.

(7.10) PROPOSITION. LCA groups are nuclear.

Proof. Let G be an LCA groups. According to (1.9), there exist a compact group K, a discrete group D and some $m = 0,1,2,\ldots$, such that G is topologically isomorphic to a subgroup of $R^m \times K \times D$. Applying (7.3) - (7.6), we see that it is enough to prove that K is nuclear. Now, K may be identified with a subgroup of some product of circles. A circle, in turn, is a quotient of the nuclear space R. That K is nuclear follows now from (7.4) - (7.6). ∎

Let G be a topological group and N a set of indices. By $c_o(N,G)$ we shall denote the group consisting of all functions $f : N \to G$ with the following property: to each $U \in N_o(G)$ there corresponds a finite subset M of N such that $F(\nu) \in U$ for all $\nu \in N \setminus M$. We make $c_o(N,G)$ into a topological group by taking as a base at zero the family of sets of the form $\{f : f(\nu) \in U \text{ for all } \nu\}$ where $U \in N_o(G)$. If G is a field of scalars, then $c_o(N,G)$ is a topological vector space over G.

(7.11) PROPOSITION. Locally convex spaces over locally compact ultrametric fields are nuclear groups.

For the terminology concerning vector spaces over ultrametric fields, we refer the reader to [68].

Proof. Let K be a locally compact ultrametric field. Every locally convex space over K is an inverse limit of Banach spaces over K. Thus, in view of (7.7), we have to show only that Banach spaces over K are nuclear groups.

Let E be a Banach space over K. We may identify E with the space $c_o(N,K)$ for some set N ([68], Corollary 2, p. 44). From standard results on the structure of locally compact fields it follows that K, as an additive topological group, may be identified with an inverse limit of discrete groups (see e.g. [103], Ch. I, §2). In other words, we may assume that K is a subgroup of some product $\prod_{i\in I} D_i$ of discrete groups. Then E may be identified with a subgroup of the group $H = c_o(N, \prod_{i\in I} D_i)$. In view of (7.5), it remains to show that H is a nuclear group.

We may treat H as a subgroup of the product

$$\prod_{\nu\in N} \left(\prod_{i\in I} D_i \right) \sim \prod_{(\nu,i)\in N\times I} D_{\nu,i}$$

where $D_{\nu,i} = D_i$. For each finite subset J of I, let

$$\pi_J : H \to D_J := \prod_{(\nu,i)\in N\times I} D_{\nu,i}$$

be the natural projection. It is clear that the family $\{\ker \pi_J\}$, where J ranges over finite subsets of I, forms a base at zero in H. Thus we may identify H with the inverse limit of the discrete groups D_J. That H is nuclear follows now from (7.3) and (7.7). ▪

(7.12) REMARK. Locally convex spaces over local fields satisfy Grothendieck's definition of nuclear spaces, based on tensor products. On the other hand, a real (or complex) topological vector space is a nuclear group if and only if it is a nuclear locally convex space (see (7.4) and (8.9)). Therefore it would be interesting to give a characterization of nuclear groups similar to some characterization of nuclear spaces. Naturally, tensor products and bilinear mappings do not make much sense for topological groups. Nevertheless, one may speak of summable and absolutely summable families of elements of abelian topological groups. In this connection, see (10.16).

(7.13) LEMMA. Let p,q be two pre-Hilbert seminorms on a vector space E, such that $\sum_{k=1}^{\infty} d_k^2(B_p,B_q) \leq 1$. Take any $u_1,\dots,u_m \in B_p$ with $m \geq 2$. If a vector $y \in E$ belongs to the set

$$\{t_1u_1 + \dots + t_mu_m : 0 \leq t_1,\dots,t_m \leq 1\},$$

then there exists a subset I of $\{1,\dots,m\}$ such that $1 \leq \operatorname{card} I \leq m-1$ and $y - \sum_{i\in I} u_i \in B_q$.

This is Lemma 4 of [10].

(7.14) LEMMA. Let p,q be two pre-Hilbert seminorms on a vector space E, such that $\sum_{k=1}^{\infty} d_k^2(B_p,B_q) \leq 1$. Let L be a subgroup of E and ψ a mapping from $K \cap 2B_q$ into some group G, such that $\psi(u+w) = \psi(u) + \psi(w)$ for all $u,w \in K \cap 2B_q$ with $u + w \in 2B_q$. Then

(a) if $u_1,\dots,u_n \in B_p$ and $u_1 + \dots + u_n \in 2B_q$, then

$$\psi(u_1 + \dots + u_n) = \psi(u_1) + \dots + \psi(u_n);$$

(b) the formula

$$\phi(\sum_{i=1}^{n} u_i) = \sum_{i=1}^{n} \psi(u_1) \qquad (u_1,\dots,u_n \in K \cap B_p)$$

defines a homomorphism $\phi : \mathrm{gp}\,(K \cap B_p) \to G$ which is identical with ψ on the set $2B_q \cap \mathrm{gp}\,(K \cap B_p)$.

Proof. To prove (a), we apply induction on n. For $n = 2$, the validity of (a) is obvious. So, assume that (a) is true for n less than some fixed integer $m = 3,4,\ldots$. We shall prove that (a) is true also for $n = m$.

Take any $u_1,\ldots,u_m \in K \cap B_p$ with

$$(1)\qquad u_1 + \ldots + u_m \in 2B_q.$$

Due to (7.13), there is a subset I of $\{1,\ldots,m\}$ with $1 \le \text{card } I \le m - 1$, such that

$$(2)\qquad \sum_{i\in I} u_i - \frac{1}{2}\sum_{i=1}^{m} u_i \in B_q.$$

Then, by (1), both $\sum_{i\in I} u_i$ and $\sum_{i\notin I} u_i$ belong to $2B_q$. Now, from our inductive assumption we obtain

$$\psi(\sum_{i=1}^{m} u_i) = \psi(\sum_{i\in I} + \sum_{i\notin I} u_i) = \psi(\sum_{i\in I} u_i) + \psi(\sum_{i\notin I} u_i)$$

$$= \sum_{i\in I} \psi(u_i) + \sum_{i\notin I} \psi(u_i) = \sum_{i=1}^{m} \psi(u_i).$$

This completes the proof of (a). To prove (b), we only have to show that ϕ is well-defined, i.e. that if $\sum_{i=1}^{n} u_i = \sum_{j=1}^{m} w_j$ for some $u_1, \ldots,u_n \in K \cap B_p$ and $w_1,\ldots,w_m \in K \cap B_p$, then $\sum_{i=1}^{n} \psi(u_i) = \sum_{j=1}^{m} \psi(u_j)$.

One has $\psi(0) = \psi(0 + 0) = \psi(0) + \psi(0)$, whence $\psi(0) = 0$. Next, for each $j = 1,\ldots,m$, one has

$$\psi(w_j) + \psi(-w_j) = \psi(w_j + (-w_j)) = \psi(0) = 0,$$

i.e. $\psi(-w_j) = -\psi(w_j)$. Thus, by (a),

$$\sum_{i=1}^{n} \psi(u_i) - \sum_{j=1}^{m} \psi(w_j) = \sum_{i=1}^{n} \psi(u_i) - \sum_{j=1}^{m} \psi(-w_j)$$

$$= \psi(\sum_{i=1}^{n} u_i + \sum_{j=1}^{m} (-w_j)) = \psi(0) = 0. \quad ■$$

(7.15) PROPOSITION. Let G,H be locally isomorphic abelian groups. If G is nuclear, so is H.

For a definition of locally isomorphic groups, see [23], Ch. III, §1, n° 1.

Proof. Choose arbitrary $U \in \mathcal{N}_o(H)$, $m = 1,2,\ldots$ and $c > 0$. We have to find a vector space E, two pre-Hilbert seminorms p,q on E with $d_k(B_p,B_q) \le ck^{-m}$ for $k = 1,2,\ldots$, a subgroup K of E and a homomorphism $\phi : K \to H$, such that $\phi(K \cap B_p) \in \mathcal{N}_o(H)$ and $\phi(K \cap B_q) \subset U$. We may assume that $c \le 1$.

Let γ be a local isomorphism of G into H. Thus γ is a homeomorphism of some $W \in \mathcal{N}_o(G)$ onto some $V \in \mathcal{N}_o(H)$, such that $\gamma(u + w) = \gamma(u) + \gamma(w)$ for any $u,w \in W$ with $u + w \in W$. We may assume that $U \subset V$. Then $U' = \gamma^{-1}(U) \in \mathcal{N}_o(G)$. Since G is a nuclear group, from (7.2) it follows that there exist a vector space E, two pre-Hilbert seminorms p,r on E such that

$$d_k(B_p,B_r) \le \tfrac{1}{2}ck^{-m-1} \qquad (k = 1,2,\ldots),$$

a subgroup K of E and a homomorphism $\psi : K \to G$, such that $\psi(K \cap B_p) \in \mathcal{N}_o(G)$ and $\psi(K \cap 2B_r) \subset U'$. By (2.15), there exists a pre-Hilbert seminorm q on E such that $d_k(B_p,B_q) \le ck^{-m}$ and $d_k(B_q,B_r) \le \frac{1}{2}k^{-1}$ for $k = 1,2,\ldots$. We have $B_p \subset B_q \subset B_r$ since $c \le 1$. Observe that

$$\sum_{k=1}^{\infty} d_k^2(B_q,B_r) \le \frac{1}{4} \sum_{k=1}^{\infty} k^{-2} < 1.$$

Let ψ' be the restriction of ψ to the set $K \cap 2B_r$. For any $u,w \in K \cap 2B_r$ with $u + w \in 2B_r$, we have

$$\begin{aligned}\psi'(u + w) &= \gamma\psi(u + w) = \gamma(\psi(u) + \psi(w)) = \gamma\psi(u) + \gamma\psi(w)\\ &= \psi'(u) + \psi'(w).\end{aligned}$$

So, according to (7.14), there is a homomorphism $\phi : K \to H$ which is identical with ψ' on the set $2B_r \cap \text{gp}\,(K \cap B_q)$. From our assumptions it now follows that

$$\phi(K \cap B_q) = \psi'(K \cap B_q) \subset \gamma\psi(K \cap B_r) \subset \gamma\psi(K \cap 2B_r) \subset \gamma(U') = U.$$

On the other hand, we have

$$\phi(K \cap B_p) = \psi'(K \cap B_p) = \gamma\psi(K \cap B_p) \in \mathcal{N}_o(H)$$

because $\phi(K \cap B_p) \in \mathcal{N}_o(G)$ and γ is a homeomorphism. ▪

(7.16) NOTE. The material of this section is new.

8. Characters of nuclear groups

(8.1) PROPOSITION. Let p,q be pre-Hilbert seminorms on a vector space E. Let K be a subgroup of E and χ a character of K with $|\chi(K \cap B_q)| \le \frac{1}{4}$. Then there is an $f \in E^{\#}$ with $\rho f_{|K} = \chi$, such that

$$\sup\{|f(u)| : u \in B_p\} \le 5 \sum_{k=1}^{\infty} k d_k(B_p,B_q),$$

provided that the right side is finite.

Proof. We may assume that

(1) $$\sum_{k=1}^{\infty} k d_k(B_p,B_q) = 1.$$

Consider the canonical diagram

$$\begin{array}{ccc} E & \xrightarrow{\ \ id\ \ } & E \\ \downarrow{\scriptstyle \psi_p} & & \downarrow{\scriptstyle \psi_q} \\ E_p & \xrightarrow{\ \ \Lambda_{pq}\ \ } & E_q \end{array}$$

Take any $u,w \in K$ with $\psi_q(u) = \psi_q(w)$. Then $\psi_q(m(u-w)) = m\psi_q(u-w) = 0$ for every m. Hence $m(u - w) \in K \cap B_q$ for every m. So, due to (1.2), we have

$$|\chi(u - w)| \le \frac{1}{m}|\chi(K \cap B_q)| \le \frac{1}{4m}$$

for every m, which implies that $\chi(u) = \chi(w)$. This proves that the formula $\kappa(\psi_q(u)) = \kappa(u)$ defines a character κ of $\psi_q(K)$. Observe that

(2) $$|\kappa(\psi_q(K) \cap B(E_q))| = |\chi(K \cap B_q)| \le \frac{1}{4}.$$

For each $u \in K$, define

$$A_u = \{h \in E_p^* : \rho h\psi_p(u) = \chi(u)\}.$$

Now, choose any $u_1,\dots,u_n \in K$. Let $M = \text{span}\,\{\psi_q(u_k)\}_{k=1}^{n} \subset E_q$ and $m = \dim M$. It follows from (2.11) that $\psi_q(B_p)$ is an absorbing subset of E_q and its Minkowski functional r is a pre-Hilbert seminorm on E_q. Therefore $M \cap \psi_q(B_p)$ and $M \cap B(E_q)$ are n-dimensional ellipsoids in M. So, by (2) and (3.15), there exists a linear functional g on M such that $\rho g = \kappa$ on $M \cap \psi_q(K)$ and

$$(3) \qquad \sup\{|g(u)| : u \in M \cap \psi_q(B_p)\} \le 5 \sum_{k=1}^{m} k d_k(M \cap \psi_q(B_p), M \cap B(E_q)).$$

Applying (2.13) and (2.8) (a), we see that

$$(4) \qquad d_k(M \cap \psi_q(B_p), M \cap B(E_q)) = d_k(M \cap B_r, M \cap B(E_q)) \le d_k(B_r, B(E_q))$$
$$= d_k(\psi_q(B_p), \psi_q(B_q)) \le d_k(B_p, B_q)$$

for every k. Now, $h' = g\Lambda_{pq}$ is a linear functional on $\Lambda_{pq}^{-1}(M)$, and from (1), (3) and (4) we get $\|h'\| \le 5$. Let $h \in E_p^*$ be any extension of h' with $\|h\| \le 5$. For each $k = 1,\dots,n$, we have

$$\rho h\psi_p(u_k) = \rho h'\psi_p(u_k) = \rho g\Lambda_{pq}\psi_p(u_k) = \rho g\psi_q(u_k) = \kappa\psi_q(u_k)$$
$$= \chi(u_k).$$

Thus $h \in 5B(E_p^*) \cap \bigcap_{k=1}^{n} A_{u_k}$.

Since the sets A_u are all weakly closed and $5B(E_p^*)$ is weakly compact, it follows that there exists some $h \in 5B(E_p^*) \cap \bigcap_{u\in K} A_u$. It remains now to take $f = h\psi_p$. ∎

(8.2) THEOREM. Let H be a subgroup of a nuclear group G. Each equicontinuous subset of $H^\frown$ is the canonical image of an equicontinuous subset of $G^\frown$.

Proof. Let X be an equicontinuous subset of $H^\frown$. There is some $W \in \mathcal{N}_o(H)$ with $X \subset W^o$. Next, there is some $U \in \mathcal{N}_o(G)$ with $U \cap H \subset W$. By (7.2), we can find a vector space E, two pre-Hilbert seminorms p, q on E with

$$(1) \qquad d_k(B_p, B_q) \le (33)^{-1}k^{-3} \qquad (k = 1,2,\dots),$$

a subgroup K of E and a homomorphism $\phi : K \to G$, such that $\phi(K \cap B_p) \in \mathcal{N}_o(G)$ and $\phi(K \cap B_q) \subset U$. Define

$$Y = \{\kappa \in G^{\wedge} : |\kappa(\phi(K \cap B_p))| \le \tfrac{1}{4}.$$

It is clear that Y is an equicontinuous subset of $G^{\wedge}$. It remains to show that each $\chi \in X$ is a restriction of some $\tilde{\chi} \in Y$.

So, take any $\chi \in X$. let $K' = \phi^{-1}(H)$. Then $\chi\phi$ is a character of K' and $|\chi\phi(K' \cap B_q)| \le |\chi(U \cap H)| \le \frac{1}{4}$. By (8.1), there is some $f \in E^{\#}$ such that $\rho f = \chi\phi$ on K' and

$$\sup\{|f(u)| : u \in B_p\} \le 5 \sum_{k=1}^{\infty} k d_k(B_p, B_q).$$

Hence, in view of (1), we have

$$\sup\{|f(u)| : u \in B_p\} \le 5\,(33)^{-1} \sum_{k=1}^{\infty} k^{-2} < \tfrac{1}{4}.$$

It is clear that the formula $\kappa(\phi(u)) = \rho f(u)$ for $u \in K$ defines a character κ of the group $\phi(K)$, with

$$|\kappa(\phi(K \cap B_p))| \le \sup\{|f(u)| : u \in B_p\} < \tfrac{1}{4}.$$

Next, the formula

$$\xi(\phi(u) + h) = \chi(\phi(u)) + \chi(h) \qquad (u \in K,\ h \in H)$$

defines a character ξ of the group $\phi(K) + H$. We have $\xi_{|\phi(K)} = \kappa$ and $\xi_{|H} = \chi$. Let us extend ξ in any way to a character $\tilde{\chi}$ of G (see (1.6)). Then $\tilde{\chi} \in Y$ and $\tilde{\chi}_{|H} = \chi$. ▪

(8.3) COROLLARY. Each subgroup of a nuclear group G is dually embedded in G. ▪

(8.4) PROPOSITION. Let p,q be pre-Hilbert seminorms on a vector space E, with $\sum_{k=1}^{\infty} k d_k(B_p, B_q) \le 1$. Let $a \in E$ and let K be a subgroup of E, such that $a \notin K + B_q$. Then there exists some $f \in E^{\#}$ with $f(K) \subset Z$, $f(a) \in [\frac{1}{4}, \frac{3}{4}] + Z$ and $\sup\{|f(u)| : u \in B_p\} \le 4$.

Proof. Define

$$Q = \{h \in E_p^{*} : h\psi_p(a) \in [\tfrac{1}{4}, \tfrac{3}{4}] + Z\}$$

and, for each $u \in K$,

$$A_u = \{h \in E_p^* : h\psi_p(u) \in Z\}.$$

We have $\psi_q(a) \notin \psi_q(K) + B(E_q)$. We may now repeat the proof of (8.1), applying (3.11) instead of (3.15), to show that $4B(E_p^*) \cap Q \cap \bigcap_{u \cap F} A_u \neq \emptyset$ for any finite subset F of K. Since Q and A_u are all weakly closed and $4B(E_p^*)$ is weakly compact, there is some $h \in 4B(E_p^*) \cap Q \cap \bigcap_{u \in K} A_u$ and it remains to take $f = h\psi_p$. ▪

(8.5) THEOREM. Nuclear groups are locally quasi-convex.

Proof. Let G be a nuclear group. Take any $U \in N_o(G)$. By (7.2), we can find a vector space E, two pre-Hilbert seminorms p,q on E with

$$(1) \qquad \sum_{k=1}^{\infty} kd_k(B_p,B_q) \leq \frac{1}{32},$$

a subgroup K of E and a homomorphism $\phi : K \to G$, such that $\phi(K \cap B_p) \in N_o(G)$ and $\phi(K \cap B_q) \subset U$. To complete the proof, it suffices to show that the quasi-convex hull of $\phi(K \cap B_q)$ is contained in U.

So, take any $g \in G \setminus U$. We are to find some $\chi \in G^\wedge$ with $|\chi(\phi(K \cap B_p))| \leq \frac{1}{4}$ and $|\chi(g)| > \frac{1}{4}$. Suppose first that $g \notin \phi(K)$. Discrete groups admit sufficiently many characters, therefore we can find a character κ of G with $\kappa_{|\phi(K)} \equiv 0$ and $\kappa(g) \neq 0$. Then $|n\kappa(g)| > \frac{1}{4}$ for a certain n and we may take $\chi = n\kappa$.

Now, suppose that $g \in \phi(K)$. Let $K' = \ker \phi$. Then $g = \phi(u)$ for some $u \notin B_q + K'$. By (1) and (8.4), there is some $f \in E^*$ with $f(K') \subset Z$, $f(u) \in [\frac{1}{4},\frac{3}{4}] + Z$ and $\sup\{|f(w)| : w \in B_p\} \leq \frac{1}{8}$. The formula $\kappa(\phi(w)) = \rho f(w)$ for $w \in K$ defines a character κ of $\phi(K)$ with $|\kappa(\phi(K \cap B_p))| \leq \frac{1}{8}$ and $|\kappa(g)| = |\kappa(\phi(u))| \geq \frac{1}{4}$. Let $\chi = \kappa$ if $|\kappa(g)| > \frac{1}{4}$ and let $\chi = 2\kappa$ if $|\chi(g)| = \frac{1}{4}$. Then $|\chi(\phi(K \cap B_p))| \leq \frac{1}{4}$ and $|\chi(g)| > \frac{1}{4}$. ▪

(8.6) COROLLARY. Closed subgroups of nuclear groups are dually closed.

Proof. Let H be a closed subgroup of a nuclear group G. Choose any $u \in G \setminus H$. We are to find some $\chi \in G^\wedge$ with $\chi_{|H} \equiv 0$ and $\chi(u) \neq 0$.

Let $\psi : G \to G/H$ be the canonical projection. Then $\psi(u) \neq 0$. Since G/H is Hausdorff, there is some $U \in N_o(G)$ with $\psi(u) \notin U$. By (7.5), the group G/H is nuclear. Therefore, by (8.5), we can find some quasi--convex set $W \in N_o(G)$ with $W \subset U$. So, there is some $\kappa \in (G/H)^{\hat{}}$ with $|\kappa\psi(u)| > \frac{1}{4}$ and we may set $\chi = \kappa\psi$. ∎

(8.7) COROLLARY. Closed subgroups of nuclear spaces are weakly closed.

This follows immediately from (8.6), (7.4) and (2.5).

(8.8) REMARK. Let K be a discrete subgroup of a nuclear space E. Then K is an at most countably generated abelian free group (see (5.6)). Since E/K admits sufficiently many continuous characters, it admits a continuous faithful unitary representation (the Hilbert sum of these characters). It was proved in [5] that if the topology of E can be defined by a family of norms, then E/K admits a faithful uniformly continuous unitary representation. Simple examples show that it is essential to assume that K is discrete.

The countable product R^{ω} of real lines does not admit a faithful uniformly continuous unitary representations. Indeed, from the proof of (4.5) it follows easily that such a representation would induce an injective continuous linear operator $\Phi : R^{\omega} \to L^{\infty}_{R}(0,1)$, which is impossible.

(8.9) PROPOSITION. Let F be a topological vector space. If F is a nuclear group, then it is a nuclear locally convex space.

Proof. From (8.5) and (2.4) we infer that F is a locally convex space. Choose any continuous seminorm s on F. By (7.2), we can find a vector space E, two pre-Hilbert seminorms p,q on E with

$$(1)\qquad d_k(B_p,B_q) \leq \tfrac{1}{3}k^{-1} \qquad (k = 1,2,\ldots),$$

a subgroup K of E and a homomorphism $\phi : K \to F$, such that $\phi(K \cap B_p) \in N_o(F)$ and $\phi(K \cap B_q) \subset B_s$. In view of (2.13), we may assume that $\operatorname{span} K = E$. The formula

$$\Phi \sum_{k=1}^{n} t_k u_k = \sum_{k=1}^{n} t_k \psi_s \phi(u_k) \qquad (t_k \in R,\ u_k \in K)$$

defines a linear operator $\Phi : E \to F_s$. To prove this, we only have to show that Φ is well-defined, i.e. that $\Sigma\, t_k\psi_s\phi(u_k) = 0$ whenever $\Sigma t_k u_k = 0$. So, choose any $u_1,\dots,u_n \in K$. Set $M = \{u_k\}_{k=1}^n$ and $H = K \cap M$. The formula $\pi(u) = \psi_s\phi(u)$ defines a homomorphism $\pi : H \to F_s$. We have $\pi(H \cap B_p) = \psi_s\phi(H \cap B_p) \subset \psi_s(B_s) = B(F_s)$, which implies that π is continuous in the usual topology on the finite dimensional space M. Let $\bar{\pi} : \bar{H} \to F_s$ be the continuous extension of π. According to (3.1), we may write $\bar{H} = \bar{H}_o \oplus \bar{H}'$ where $\bar{H}_o$ is a closed linear subspace of M and $\bar{H}'$ is a free discrete group. We can therefore extend $\bar{\pi}$ to a continuous homomorphism $\sigma : M \to F_s$. Being additive and continuous, σ is a linear mapping. Thus

$$\sum_{k=1}^{n} t_k\psi_s\phi(u_k) = \sum_{k=1}^{n} t_k\sigma(u_k) = \sigma(\sum_{k=1}^{n} t_k u_k) = \sigma(0) = 0,$$

which proves that Φ is well-defined. Set $X = \operatorname{conv}(K \cap B_p)$ and $Y = \operatorname{conv}(K \cap B_q)$. From (1) we get

$$\sum_{k=1}^{\infty} d_k^2(B_p,B_q) \le \frac{1}{9}\sum_{k=1}^{\infty} k^{-2} = \frac{\pi^2}{54} < \frac{1}{4}.$$

Hence, in virtue of (3.20),

(2) $$d_k(X,Y) \le 2d_k(B_p,B_q) \qquad (k = 1,2,\dots).$$

It is not difficult to see that $\psi_s^{-1}(\Phi(Y)) \subset B_s$ and $\psi_s^{-1}(\Phi(X)) \supset \phi(K \cap B_p)$, whence $\psi_s^{-1}(\Phi(X)) \in N_o(F)$. Finally, from (1), (2) and (2.8) we obtain

$$d_k(\psi_s^{-1}(\Phi(X)),\psi_s^{-1}(\Phi(Y))) \le d_k(X,Y) < k^{-1} \qquad (k = 1,2,\dots).$$

This proves that F is a nuclear space because s was an arbitrary continuous seminorm on F. ∎

(8.10) REMARKS. Let E,F be two normed spaces and let $\mathfrak{C}(E,F)$ be the family of all bounded linear operators $\Phi : E \to F$ which satisfy the following condition: if K is a closed subgroup of F, then $\Phi^{-1}(K)$ is a weakly closed subgroup of E (or, which is the same, if K is a subgroup of E, then Φ maps the weak closure of K in E into the closure of $\Phi(K)$ in F). Proposition (8.4) says that if E, F are unitary and $\sum_{k=1}^{\infty} kd_k(\Phi) < \infty$, then $\Phi \in \mathfrak{C}(E,F)$. On the other hand, from the proof of Lemma 6 in [4] it follows that if E,F are arbitrary normed spaces and $\Phi \in \mathfrak{G}(E,F)$, then $kv_k(\Phi)$ remains bounded as $k \to \infty$.

Let $\Psi : E' \to E$ and $\chi : F \to F'$ be arbitrary bounded linear operators. It is clear that if $\Phi \in \mathfrak{G}(E,F)$, then $\Phi\Psi \in (E',F)$ and $\chi\Phi \in \mathfrak{G}(E,F')$. However, it is not known whether the sum of two operators from $\mathfrak{G}(E,F)$ belongs to $\mathfrak{G}(E,F)$.

The situation described above is typical in the following sense. A theorem on additive subgroups of nuclear spaces leads to a certain statement on linear operators between unitary spaces which, in turn, is reduced to a statement on lattices and ellipsoids in R^n (cf. the remarks at the end of section 3). This gives rise to a certain class $\mathfrak{A}$ of operators acting in Banach spaces; $\mathfrak{A}$ is usually closed with respect to composition with bounded operators but probably not with respect to addition of operators. If $d_k(\Phi)$ are small enough (say, $d_k(\Phi) \sim k^{-1}$ or $\sim k^{-2}$), then $\Phi \in \mathfrak{A}$. On the other hand, if $\Phi \in \mathfrak{A}$, then $v_k(\Phi)$ cannot be too large.

Most important examples of such classes of operators are connected with (8.4) (the class $\mathfrak{G}$ considered above), with (8.1) and with (12.2). Very interesting (and probably difficult) problems of this appear in connection with the considerations of section 6. See also (10.18).

(8.11) NOTE. The material of this section is mostly new. For subgroups and quotient groups of nuclear spaces, the results of this section were proved in [7] and [8]. Proposition (8.1) is a strengthening of Lemma 1.5 of [8].

9. Nuclear vector groups

(9.1) DEFINITION. Let F be a vector space and τ a topology on F such that F_τ is an additive topological group. We say that F_τ is a locally convex vector group if it is separated and has a base at zero consisiting of symmetric, convex sets.

This notion was introduced by D.A. Raikov in [78], p. 301. Obthat, for each $\lambda \in R$, the mapping $u \to \lambda u$ of F_τ into itself is continuous. Observe also that F_τ is a topological vector space if and only if it has a base at zero consisting of absorbing sets; then it is a locally convex space.

(9.2) DEFINITION. A locally convex vector group F is called a <u>nuclear vector group</u> if to each symmetric, convex $U \in \mathcal{N}_o(F)$ there corresponds a symmetric, convex $W \in \mathcal{N}_o(F)$ such that

$$d_k(W,U) \le k^{-1} \qquad (k = 1,2,\ldots).$$

(9.3) PROPOSITION. Let F be a nuclear vector group. Take any $m = 1,2,\ldots$ and $c > 0$. To each radial $U \in \mathcal{N}_o(F)$ there corresponds some symmetric, convex $W \in \mathcal{N}_o(F)$ with

$$d_k(W,U) \le ck^{-m} \qquad (k = 1,2,\ldots).$$

The proof is similar to that of (2.17).

(9.4) PROPOSITION. Every nuclear vector group is a nuclear group.

This is an immediate consequence of (7.1) and (9.3) (see the proof of (7.4)).

(9.5) PROPOSITION. The completion of a nuclear vector group is a nuclear vector group.

Proof. Let F be a nuclear vector group. We introduce multiplication by real numbers in $\tilde{F}$ in the following way. Let $\tilde{f} \in \tilde{F}$ and $t \in R$. There is a generalized sequence (f_σ) in F, converging to $\tilde{f}$. Then (tf_σ) is a Cauchy sequence in F. We set $t\tilde{f} = \lim tf_\sigma$. It is easy to see that $\tilde{F}$ with the multiplication thus defined is a vector space.

Let $\{U_i\}_{i\in I}$ be a base at zero in F, consisting of symmetric, convex sets. Then $\{\overline{U}_i\}_{i\in I}$ is a base at zero in $\tilde{F}$. It is easy to see that the sets $\overline{U}_i$ are symmetric and convex. Choose any $i \in I$. There is some $j \in I$ such that

$$d_k(U_j,U_i) < k^{-1} \qquad (k = 1,2,\ldots).$$

For any fixed k, we can find a linear subspace L of F with $\dim L < k$ and $U_j \subset k^{-1}U_i + L$. For each $\varepsilon > 0$, we have $\overline{U}_j \subset U_j + \varepsilon\overline{U}_j$, whence

$$\overline{U}_j \subset k^{-1}U_i + \varepsilon\overline{U}_j + L \subset k^{-1}\overline{U}_i + L = (k^{-1} + \varepsilon)\overline{U}_i + L.$$

Therefore

$$d_k(\bar{U}_j,\bar{U}_i) \le k^{-1} \qquad (k = 1,2,\ldots),$$

which proves that $\tilde{F}$ is a nuclear vector group. ▪

Let G be a topological group. By $w_o(G)$ we shall denote the weight of G at zero, i.e. the least cardinal number m such that G has a base at zero with cardinality m.

(9.6) THEOREM. Let G be a nuclear group. Then there exist a nuclear vector group F with $w_o(F) = w_o(G)$, a subgroup H of F and a closed subgroup Q of H, such that G is topologically isomorphic to H/Q.

Proof. Choose a base $\mathcal{B}$ at zero in G with $\operatorname{card}\mathcal{B} = w_o(G)$. Due to (8.5), we may assume that $\mathcal{B}$ consists of quasi-convex sets. Let $R^{G^\wedge}$ be the vector space of all real-valued functions on $G^\wedge$. For each $U \in \mathcal{B}$, let

$$X_U = \{\xi \in R^{G^\wedge} : \exists_{g\in U} \ \forall_{\chi\in U^o} \ \xi(\chi) = \chi(g)\}$$

and let $Y_U = \operatorname{conv} X_U$. If $V \in \mathcal{B}$ and $V + V \subset U$, then it is clear that $Y_V + Y_V \subset Y_U$. It is also clear that $Y_{-U} = -Y_U$ because both U and Y are symmetric sets. Thus, according to (1.12), there is a unique topology τ on $R^{G^\wedge}$ such that $F := (R^{G^\wedge},\tau)$ is a topological group for which $\{Y_U\}_{U\in\mathcal{B}}$ is a base at zero.

Consider the homomorphism $\sigma : G \to T^{G^\wedge}$ given by the formula

$$\sigma(g)(\chi) = \chi(g) \qquad (g \in G, \ \chi \in G^\wedge).$$

It follows from (8.5) that $G^\wedge$ separates points of G. Therefore σ is injective. Let $\rho_G : R^{G^\wedge} \to T^{G^\wedge}$ be the canonical projection given by the formula

$$\rho_G(\xi)(\chi) = \rho(\xi(\chi)) \qquad (\xi \in R^{G^\wedge}, \ \chi \in G^\wedge).$$

Set $H = \rho_G^{-1}(\sigma(G))$ and $Q = H \cap \rho_G^{-1}(0)$. We shall prove that H/G with the topology induced from F is topologically isomorphic to G. Consider the canonical diagram

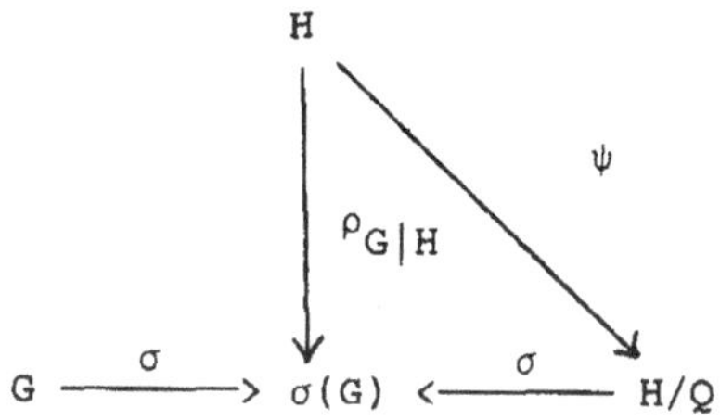

Here σ and π are both algebraical isomorphisms. We shall prove that $\sigma^{-1}\pi$ is a topological isomorphism.

Choose any $U \in \mathcal{B}$. We are going to show that

(1) $$\rho_G(H \cap Y_U) = \sigma(U).$$

We begin with the inclusion

(2) $$\rho_G(H \cap Y_U) \supset \sigma(U).$$

So, take any $\xi \in \sigma(U)$. We have $\xi = \sigma(g)$ for some $g \in U$. We may treat ξ as an element of $R^{G^{\frown}}$. Then $\xi(\chi) = \sigma(g)(\chi) = \chi(g)$ for each $\chi \in G^{\frown}$, which implies that $\xi \in X_U$. Next, we have $\rho_G(\xi) = \xi = \sigma(g) \in \sigma(G)$, whence $\xi \in H$. Thus $\xi \in H \cap Y_U$ and $\xi = \rho_G(\xi) \in \rho_G(H \cap Y_U)$, which proves (2).

To prove the opposite inclusion, choose any $\xi \in H \cap Y_U$. Since $Y_U = \operatorname{conv} X_U$, and X_U is symmetric, we may write $\xi = t_1\xi_1 + \ldots + t_n\xi_n$ for some $\xi_1,\ldots,\xi_n \in X_U$ and some $t_1,\ldots,t_n > 0$ with $t_1 + \ldots + t_n = 1$. According to the definition of X_U, for each $k = 1,\ldots,n$, there is some $g_k \in U$ with $\xi_k(\chi) = \chi(g_k)$ for all $\chi \in U^{\circ}$. Hence $|\xi_k(\chi)| \leqq \frac{1}{4}$ for all $x \in U^{\circ}$ and $k = 1,\ldots,n$. From this we get

(3) $$|\xi(\chi)| \leqq t_1|\xi_1(\chi)| + \ldots + t_n|\xi_n(\chi)| \leqq \frac{1}{4} \qquad (\chi \in U^{\circ}).$$

Now, since $\xi \in H$, we have $\rho_G(\xi) = \sigma(g)$ for some $g \in G$. By (3), for each $\chi \in U^{\circ}$, we have

$$|\chi(g)| = |\sigma(g)(\chi)| = |\rho_G(\xi)(\chi)| = |\rho(\xi(\chi))| \leqq |\xi(\chi)| \leqq \frac{1}{4}.$$

This implies that $g \in U$ because we have assumed U to be quasi-convex. Hence $\rho_G(\xi) = \sigma(g) \in \sigma(U)$, which yields $\rho_G(H \cap Y_U) \subset \sigma(U)$ since $\xi \in H \cap Y_U$ was arbitrary. From this and (2) we obtain (1).

Now, (1) may be written as $\sigma^{-1}\pi\psi(H \cap Y_U) = U$. Since this holds for all $U \in \mathcal{B}$, we get $\{\sigma^{-1}\pi\psi(H \cap Y_U)\}_{U\in\mathcal{B}} = \mathcal{B}$. But $\{\psi(H \cap Y_U)\}_{U\in\mathcal{B}}$ is a base at zero in H/Q, and $\mathcal{B}$ is such a base in G. Hence $\sigma^{-1}\pi$ is a topological isomorphism.

To complete the proof, it remains to show that F is a nuclear vector group. So, take any $U \in N_o(F)$. There is some $V \in \mathcal{B}$ with $Y_V \subset U$. By (7.2), there is a vector space E, two pre-Hilbert seminorms p,q on E with

$$d_k(B_p,B_q) \le 10^{-2}k^{-4} \qquad (k = 1,2,\ldots),$$

a subgroup K of E and a homomorphism $\phi : K \to G$, such that $\phi(K \cap B_p) \in N_o(G)$ and $\phi(K \cap B_p) \subset \frac{1}{2}V$. Due to (2.15), we can find a pre-Hilbert seminorm r on E with $p \ge r \ge q$,

$$(4) \qquad d_k(B_r,B_q) < \frac{1}{33}k^{-3} \qquad (k = 1,2,\ldots),$$

$$(5) \qquad d_k(B_p,B_r) < \frac{1}{3}k^{-1} \qquad (k = 1,2,\ldots).$$

Now, we shall construct a linear operator $\Phi : E \to R^{G^\frown}$. Take any $\chi \in V^o$. Then, by (1.2),

$$|\chi(\tfrac{1}{2}V)| \le \tfrac{1}{2}|\chi(V)| \le \tfrac{1}{2}\cdot\tfrac{1}{4} < \tfrac{1}{4},$$

so $|\chi\phi(K \cap B_p)| < \frac{1}{4}$. In virtue of (8.1), there is some $f_\chi \in E^\#$ with $\rho f_\chi = \chi\phi$ on K and

$$\sup\{|f_\chi(u)| : u \in B_r\} \le 5 \sum_{k=1}^{\infty} k d_k(B_r,B_q).$$

Hence, by (4), we get

$$(6) \qquad \sup\{|f_\chi(u)|: u \in B_r\} < \frac{5}{33}\sum_{k=1}^{\infty} k^{-2} < \frac{1}{4}.$$

For each $u \in E$, define

$$(\Phi u)(\chi) = \begin{cases} f_\chi(u) & \text{if } \chi \in V^o, \\ 0 & \text{if } \chi \notin V^o. \end{cases}$$

Let $P : R^{G^\wedge} \to R^{G^\wedge}$ be the projection given by the formula

$$(P\zeta)(\chi) = \begin{cases} \zeta(\chi) & \text{if} \quad \chi \in V^o, \\ 0 & \text{if} \quad \chi \notin V^o. \end{cases}$$

We shall prove that

(7) $\quad P(H) \cap \Phi(B_r) \subset P(X_V)$.

So, take any $\xi \in P(H) \cap \Phi(B_r)$. Then $\xi = P\zeta$ for some $\zeta \in H$ and $\xi = \Phi u$ for some $u \in B_r$. Since $\zeta \in H$, there is some $g \in G$ with $\rho_G(\zeta) = \sigma(g)$. For each $\chi \in V^o$, we have

(8) $$\chi(g) = \sigma(g)(\chi) = \rho_G(\zeta)(\chi) = \rho(\zeta(\chi)) = \rho(P\zeta(\chi)) = \rho(\xi(\chi))$$
$$= \rho(\Phi u(\chi)) = \rho f_\chi(u).$$

Since $u \in B_r$, from (6) it follows that

$$|\chi(g)| = |\rho f_\chi(u)| \leq |f_\chi(u) < \tfrac{1}{4} \qquad (\chi \in V^o).$$

Thus $\zeta \in X_V$ and $\xi = P\zeta \in P(X_V)$, which proves (7).

Now, since $\phi(K \cap B_p) \in N_o(G)$, there is some $W \in \mathcal{B}$ with $W \subset \phi(K \cap B_p)$. We shall prove that

(9) $\quad P(X_W) \subset P(H) \cap \Phi(B_p)$.

Choose any $\xi \in P(X_W)$. There is some $\zeta \in X_W$ with $\xi = P\zeta$. Next, there is some $g \in W$ with $\zeta(\chi) = \chi(g)$ for all $\chi \in W^o$. Finally, we have $g = \phi(u)$ for some $u \in K \cap B_p$. Now, take any $\chi \in V^o$. Since $B_p \subset B_r$, we have $u \in B_r$, and from (6) we obtain $|f_\chi(u)| < \frac{1}{4}$. Then

$$\xi(\chi) = \zeta(\chi) = \chi(g) = \chi\phi(u) = \rho f_\chi(u) = f_\chi(u) = (\Phi u)(\chi)$$

(observe that $V^o \subset W^o$ because $W \subset V$). For $\chi \notin V^o$, we have $\xi(\chi) = (\Phi u)(\chi) = 0$. Hence $\xi = \Phi u \in \Phi(B_p)$, which implies that

(10) $\quad P(X_W) \subset \Phi(B_p)$.

Let $\eta \in R^{G^\frown}$ be given by the formula $\eta(\chi) = \chi(g)$ for $\chi \in G^\frown$. Then $\rho_G(\eta)(\chi) = \chi(g) = \sigma(g)(\chi)$ for every χ, i.e. $\rho_G(\eta) = \sigma(g)$. Thus $\eta \in H$. Moreover, $P\eta = \xi$, which implies that $P(X_W) \subset P(H)$. From this and (10) we obtain (9).

Set $M = \Phi(E)$. The Minkowski functionals of $\Phi(B_p)$ and $\Phi(B_r)$ are pre-Hilbert seminorms on M (see (2.11)). From (2.8) (a) it follows that

(11) $$d_k(\Phi(B_p),\Phi(B_r)) \leq d_k(B_p,B_r) \qquad (k = 1,2,\ldots).$$

hence, by (5),

$$\sum_{k=1}^{\infty} d_k^2(\Phi(B_p),\Phi(B_r)) \leq \frac{1}{9} \sum_{k=1}^{\infty} k^{-2} < \frac{1}{4}.$$

Now, (3.20) implies that

(12) $$d_k(\operatorname{conv}(P(H) \cap \Phi(B_p)), \operatorname{conv}(P(H) \cap \Phi(B_r))) \leq 2d_k(\Phi(B_p),\Phi(B_r)) \qquad (k = 1,2,\ldots).$$

We have $Y_W \subset P^{-1}(P(Y_W))$ and $Y_V = P^{-1}(P(Y_V))$. Finally, by (2.8) (b), (7), (9), (11), (12) and (5), we obtain

$$\begin{aligned} d_k(Y_W,Y_V) &\leq d_k(P^{-1}(P(Y_W)),P^{-1}(P(Y_V))) = d_k(P(Y_W),P(Y_V)) \\ &= d_k(P(\operatorname{conv} X_W),P(\operatorname{conv} X_V)) = d_k(\operatorname{conv} P(X_W),\operatorname{conv} P(X_V)) \\ &\leq d_k(\operatorname{conv}(P(H) \cap \Phi(B_p)),\operatorname{conv}(P(H) \cap \Phi(B_r))) \leq \tfrac{2}{3}k^{-1} < k^{-1} \end{aligned}$$

for every k. Thus F is a nuclear vector group.

Since $\{W_V\}_{V\in\mathcal{B}}$ is a base at zero for F, we have

$$w_o(F) \leq \operatorname{card} \{Y_V\}_{V\in\mathcal{B}} \leq \operatorname{card} \mathcal{B} = w_o(G).$$

On the other hand, $w_o(G) = w_o(H/Q) \leq w_o(H) \leq w_o(F)$. ∎

(9.7) PROPOSITION. Let G be a metrizable nuclear group. Then there exist a metrizable and complete nuclear vector group F, a closed subgroup H of F and a closed subgroup Q of H, such that the completion of G is topologically isomorpic to H/Q.

Proof. By (9.6), there exist a metrizable nuclear vector group F', a subgroup H' of F' and a closed subgroup Q' of H', such that G is topologically isomorphic to H'/Q'. Let F be the completion of F' and let H and Q be the closures in F of H' and Q', respectively. From (1.10) it now follows that the completion of G is topologically isomorphic to H/Q. Finally, F is a nuclear vector group, due to (9.5). ∎

(9.8) THEOREM. The completion of a metrizable nuclear group is a nuclear group.

This follows immediately from (9.7), (9.4) and (7.5).

(9.9) REMARK. It follows from (9.6), (9.4) and (7.5) that we might define nuclear groups as Hausdorff quotients of subgroups of nuclear vector groups. Such a definition is much simpler and allows us to omit the long and complicated proof of (9.6). However, it has also some disadvantages. Firstly, it is not "intrinsic". Secondly, applying this definition, we would encounter difficulties in section 7, with proving that groups locally isomorphic to nuclear groups are nuclear. Also, which is more important, we would have troubles in section 16 with showing that the dual group of a metrizable nuclear group is nuclear. In fact, we would have to repeat there the argument from the proof of (9.6).

(9.10) NOTE. The material of this section is new.

10. The Lévy-Steinitz theorem

Let $\sum_{i=1}^{\infty} g_i$ be a convergent series in a Hausdorff abelian group G. Its sum will be denoted simply by $\sum_{i=1}^{\infty} g_i$, that is,

$$\sum_{i=1}^{\infty} g_i = \lim_{j\to\infty} \sum_{i=1}^{j} g_i .$$

To simplify the notation, we shall often write Σg_i instead of $\sum_{i=1}^{\infty} g_i$ to denote both the series and its sum; this should not lead to misunderstandings.

The set of sums of the series Σg_i, denoted by $\mathcal{S}(\Sigma g_i;G)$ is defined in the following way: a point $s \in G$ belongs to $\mathcal{S}(\Sigma g_i;G)$

if there is a permutation π of indices such that the series $\Sigma g_{\pi(i)}$ converges to s.

(10.1) REMARK. It may happen that $\mathcal{S}(\Sigma g_i;\tilde{G}) \neq \mathcal{S}(\Sigma g_i;G)$. For example, denoting the group of rational numbers by Q, we have

$$\mathcal{S}\left(\sum_{i=1}^{\infty} \frac{(-1)^i}{i};Q\right) = Q \quad \text{and} \quad \mathcal{S}\left(\sum_{i=1}^{\infty} \frac{(-1)^i}{i};R\right) = R.$$

If the meaning of G is clear, we shall write simply $\mathcal{S}(\Sigma g_i)$ instead of $\mathcal{S}(\Sigma g_i;G)$.

The Lévy-Steinitz theorem says that the set of sums of a convergent series in R^n is an affine subspace (cf. also (10.11)). This theorem can be generalized to certain infinite dimensional spaces; see the remarks following (10.2). For series in Banach spaces, see (10.12) and (10.13).

On p. 61 of his book [89], S. Ulam posed the following problem: in connection with the Lévy-Steinitz theorem, he and Garrett Birkhoff had noticed that the following assertion is true:

(*) the set of sums of a convergent series in a compact group G is a coset modulo a certain subgroup of G;

does a similar result hold for more general, non-compact topological groups? This section is an attempt to answer the question; the main result is (10.3).

(10.2) NOTE. Ulam and Birkhoff did not leave, as it seems, any indications and one does not very well know how they had shown (*). For compact metrizable groups, (*) follows from (10.3). In this particular case, the proof of (10.3) can be much simplified, for it suffices to apply (10.9) and to make use of the fact that the set of sums of a convergent series in the countable product of real lines is a linear manifold (see below). Without the metrizability assumption, (*) is false; see (10.14).

Let E be a real topological vector space. By a <u>linear manifold</u> in E we mean a set of the form $u + M$ where $u \in E$ and M is a linear subspace of E. et us consider the following assertion:

(**) the set of sums of every convergent series in E is a closed linear manifold.

If E is the locally convex direct sum of an arbitrary number of real lines, the validity of (**) follows immediately from the Lévy-Steinitz theorem for R^n because every convergent series in E lies in a finite dimensional subspace.

Troyanski [88] proved that (**) is true if E is the countable product of real lines. His result had been obtained earlier by Wald [100] and was rediscovered later by Katznelson and McGehee [51]. See also Halperin [34]. The validity of (**) for some other nuclear spaces can be deduced from the results of Bárány [16], as well as from those of Beck [19] and Pecherskii [74]; see (10.18). By modifying Bárány's proof, the author showed in [10] that (**) is satisfied for every metrizable nuclear space E; cf. also (10.8), (10.11) and (10.14).

The aim of this section is to prove the following fact:

(10.3) THEOREM. Let Σg_i be a convergent series in a metrizable nuclear group G. Then $P := \mathcal{S}(\Sigma g_i) - \Sigma g_i$ is subgroup of G (not necessarily closed). If G is complete, P is a continuous homomorphic image of a nuclear Fréchet space.

The assumption of metrizability is essential; see (10.14).

Let us begin with some definitions. Let Σg_i be a convergent series in a Hausdorff abelian group G. By $\mathfrak{C}(\Sigma g_i;G)$ we shall denote the set defined in the following way: a point $s \in G$ belongs to $\mathfrak{C}(\Sigma g_i;G)$ if there exist a permutation π of indices and a sequence $j_1 < j_2 < \ldots$, such that

$$s = \lim_{n\to\infty} \sum_{i=1}^{j_n} g_{\pi(i)}.$$

For each $m = 1,2,\ldots$, let A_m be the closure in G of the set of all points of the form $\sum_{i\in I} g_i$ where I is a finite subset of $\{m,m+1,\ldots\}$. We define $\mathfrak{A}(\Sigma g_i;G) = \bigcap_{m=1}^{\infty} A_m$.

Remarks analogous to (10.1) are applied to $\mathfrak{C}(\Sigma g_i;G)$ and $\mathfrak{A}(\Sigma g_i;G)$, as well. When there is no risk of misunderstanding, we shall briefly write $\mathfrak{C}(\Sigma g_i)$ and $\mathfrak{A}(\Sigma g_i)$.

(10.4) PROPOSITION. Let Σg_i be a convergent series in a Hausdorff abelian group G. Then $\mathfrak{A}(\Sigma g_i)$ is a closed subgroup G. Moreover,

$\mathfrak{A}(\Sigma g_i) + \Sigma g_i = \mathfrak{C}(\Sigma g_i)$.

This is a well-known fact; see e.g. [23], Ch. III, §5, Exercise 3. For the first time, it appears in Wald's paper [101]. For series in R^n, it was known to Steinitz [87].

(10.5) LEMMA. Let $D \subset E \subset F$ be three o-symmetric n-dimensional closed ellipsoids in R^n with

$$\sum_{k=1}^{n} d_k^2(D,E) \leq 1 \qquad \text{and} \qquad \sum_{k=1}^{n} d_k^2(E,F) \leq \frac{1}{4}.$$

Let $u_1,\ldots,u_m \in D$ and $a \in E$ be such that $a + \sum_{i=1}^{m} u_i \in E$. Then there exists a permutation σ of $\{1,\ldots,m\}$ such that

$$a + \sum_{i=1}^{j} u_{\sigma(i)} \in F \qquad (j = 1,\ldots,m).$$

This is Lemma 6 of [10].

(10.6) LEMMA. If Σu_i is a convergent series in a metrizable nuclear vector group, then $\mathfrak{S}(\Sigma u_i) = \mathfrak{C}(\Sigma u_i)$.

Proof. We only have to show that $\mathfrak{C}(\Sigma u_i) \subset \mathfrak{S}(\Sigma u_i)$, the opposite inclusion being trivial. So, take any $s \in \mathfrak{C}(\Sigma u_i)$. There are a permutation π of indices and a sequence $j_1 < j_2 < \ldots$, such that

$$s = \lim_{n\to\infty} \sum_{i=1}^{j_n} u_{\pi(i)}.$$

Let us denote our nuclear vector group by F. It follows easily from (9.3) and (2.14) that we can find a fundamental sequence $U_1 \supset U_2 \supset \ldots$ of convex neighbourhoods of zero in F, such that, for every n, the Minkowski functional p_n of U_n is a pre-Hilbert seminorm on $M_n := \operatorname{span} U_n$, and

(1) $\qquad U_n = \{u \in M_n : p_n(u) \leq 1\}$,

(2) $\qquad \sum_{k=1}^{\infty} d_k^2(U_{n+1}, U_n \cap M_{n+1}) \leq \frac{1}{4}$.

Increasing the indices j_n, if need be, we may assume that, for every n,

$$(3) \qquad s - \sum_{i=1}^{j_n} u_{\pi(i)} \in U_{n+2},$$

$$(4) \qquad u_{\pi(i)} \in U_{n+2} \quad \text{for all} \quad i > j_n.$$

Let us now fix an arbitrary index n. Replacing n by $n+1$ in (2), we get

$$(5) \qquad \sum_{k=1}^{\infty} d_k^2(U_{n+2}, U_{n+1} \cap M_{n+2}) \leq \frac{1}{4} < 1.$$

Similarly, replacing n by $n+1$ in (3), we get

$$(6) \qquad s - \sum_{i=1}^{j_{n+1}} u_{\pi(i)} \in U_{n+2}$$

because $U_{n+3} \subset U_{n+2}$. Let L be the linear subspace of F spanned over the vector $s - \sum_{i=1}^{j_n} u_{\pi(i)}$ and the vectors u_i for $i = j_n+1, \ldots, j_{n+1}$. Set $l = \dim L$. From (3) and (4) we get $L \subset M_{n+2}$. Therefore, replacing n by $n+1$ and $n+2$ in (1), we see that $U_n \cap L, U_{n+1} \cap L$ and $U_{n+2} \cap L$ are o-symmetric closed l-dimensional ellipsoids in L. In virtue of (2.13), from (5) and (2) we obtain

$$\sum_{k=1}^{l} d_k^2(U_{n+2} \cap L, U_{n+1} \cap L) \leq 1, \qquad \sum_{k=1}^{l} d_k^2(U_{n+1} \cap L, U_n \cap L) \leq \frac{1}{4}.$$

Thus, by (4), (3) and (6), from (10.5) we infer that there exists a permutation σ_n of the set $\{\pi(j_n+1), \ldots, \pi(j_{n+1})\}$, such that

$$(7) \qquad s - \sum_{i=1}^{j_n} u_{\pi(i)} - \sum_{i=j_n+1}^{j} u_{\sigma_n \pi(i)} \in U_n \qquad (j_n + 1 \leq j \leq j_{n+1}).$$

Let ρ be the permutation of positive integers given by $\rho(i) = \sigma_n \pi(i)$ when $j_n + 1 \leq 1 \leq j_{n+1}$ and by $\rho(i) = i$ when $i < j_1$. Since (7) holds for every n, we may write

$$s - \sum_{i=1}^{j} u_{\rho(i)} \in U_n \qquad (j > j_n;\quad n = 1,2,\ldots).$$

This means that the series $\Sigma\, u_{\rho(i)}$ converges to s. Thus $s \in \mathcal{S}(\Sigma u_i)$. ∎

Let F be a nuclear vector group. The set $\bigcap\{\text{span } U : U \in \mathcal{N}_0(F)\}$ is a closed linear subspace of F; we shall denote it by F_0. Notice that F_0 is a nuclear space.

(10.7) **LEMMA.** Let Σu_i be a convergent series in a nuclear vector group F. Then $\mathcal{A}(\Sigma u_i)$ is a closed linear subspace of F contained in F_0.

Proof. For each $m = 1,2,\ldots,$ let A_m be the closure in F of the set of all points of the form $\sum_{i\in I} u_i$ where I is a finite subset of $\{m,m+1,\ldots\}$. Denote $\mathcal{A} = \mathcal{A}(\Sigma u_i)$; then $\mathcal{A} = \bigcap_{m=1}^{\infty} A_m$. Since $u_i \to 0$, to each $U \in \mathcal{N}_0(F)$ there corresponds an index m such that $u_i \in U$ for $i \geq m$; hence $A_m \subset \overline{\text{span}}\, U = \text{span } U$. This implies that $\mathcal{A} \subset F_0$. Since $\mathcal{A}$ is, by (10.4), a closed subgroup of F, we only have to prove that $\mathcal{A}$ is radial.

Choose arbitrary $s \in \mathcal{A}$ and $\vartheta \in (0,1)$. It is to be shown that $\vartheta s \in \mathcal{A}$. So, take any $m = 1,2,\ldots$ and any $U \in \mathcal{N}_0(F)$. Due to (9.3) and (2.14), we can find some $W \in \mathcal{N}_0(F)$ and two pre-Hilbert seminorms p,q on span W, such that $W \cap B_p \cap B_q \cap U$ and

(1) $$\sum_{k=1}^{\infty} d_k^2(B_p,B_q) \leq 1.$$

Since $B_p \in \mathcal{N}_0(F)$ and $u_i \to 0$, there is an index $n \geq m$ such that

(2) $$u_i \in B_p \quad \text{for all } i \geq n.$$

Next, since $U \in \mathcal{N}_0(F)$ and $s \in A_n$, there is a finite subset I of $\{n,n+1,\ldots\}$ such that

(3) $$s \in \sum_{i\in I} u_i + U.$$

From (1), (2) and (7.13) it follows that there is a subset J of I such that

$$\vartheta \sum_{i \in I} u_i \in B_q + \sum_{i \in J} u_i .$$

From this and (3) we derive

$$\vartheta s \in \vartheta U + \vartheta \sum_{i \in I} u_i \subset (1 + \vartheta) + \sum_{i \in J} u_i .$$

Since $U \in N_o(F)$ was arbitrary, it follows that $\vartheta s \in A_n$. But $n \geq m$, therefore $A_n \subset A_m$ and, consequently, $\vartheta s \in A_m$. This proves that $\vartheta s \in A$ because m was arbitrary. ▪

(10.8) COROLLARY. Let Σg_i be a convergent series in a metrizable nuclear vector group F. Then $\mathcal{S}(\Sigma g_i) - \Sigma g_i$ is a closed linear subspace of F contained in F_o.

This is a direct consequence of (10.4), (10.6) and (10.7). The assumption of metrizability is essential; see (10.14).

(10.9) LEMMA. Let X be a set of indices and let R^X denote the space of all real-valued functions on X, endowed with the topology of pointwise convergence. Similarly, let T^X denote the group of all functions $g : X \to T$ endowed with the topology of pointwise convergence and let $\psi : R^X \to T^X$ be the natural projection given by

$$\psi(u)(x) = \rho(u(x)) \qquad (u \in R^X; \quad x \in X).$$

For each convergent series Σu_i in R^X, one has

$$\mathcal{S}(\Sigma \psi(u_i); T^X) = \psi(\mathcal{S}(\Sigma u_i; R^X)).$$

Proof. The inclusion $\psi(\mathcal{S}(\Sigma u_i)) \subset \mathcal{S}(\Sigma \psi(u_i))$ is trivial. To prove the opposite one, take any $g \in \mathcal{S}(\Sigma \psi(u_i))$. There is a permutation π of positive integers such that

$$(1) \qquad \sum_{i=1}^{j} \psi(u_{\pi(i)}) \to g \qquad \text{as} \quad j \to \infty.$$

Fix an arbitrary $x \in X$. From (1) we get

$$\sum_{i=1}^{j} \psi(u_{\pi(i)})(x) \to g(x) \qquad \text{as} \quad j \to \infty.$$

which can be written as

(2) $\rho(\sum_{i=1}^{j} u_{\pi(i)}(x)) \to g(x) \qquad \text{as} \quad j \to \infty.$

Since (u_i) is a null sequence in R^X, we have $u_i(x) \to 0$ as $i \to \infty$. From this and (2) it follows that there is an integer $k(x)$ such that

(3) $\sum_{i=1}^{j} u_{\pi(i)}(x) \to g(x) + k(x) \qquad \text{as} \quad j \to \infty.$

Since (3) holds for each $x \in X$, the series $\Sigma u_{\pi(i)}$ converges to the function $g + k$ in R^X. Thus $g + k \in \mathcal{S}(\Sigma u_i)$. We have

$$\psi(g + k) = \psi(g) + \psi(k) = g + 0 = g,$$

which proves that $g \in \psi(\mathcal{S}(\Sigma u_i))$. ■

The proof of (10.3), given below, is, in fact, a modification of the proof of (10.9). The differences may bear a technical character only.

Proof of (10.3). It is clear that

$$\mathcal{S}(\Sigma g_i;G) = G \cap \mathcal{S}(\Sigma g_i;\tilde{G}),$$

therefore we may assume that G is complete. By (9.7), we may write $G = H/K$ where H is a closed subgroup of some metrizable and complete nuclear vector group F and K is a closed subgroup of S. Since H/K may be identified with a closed subgroup of F/K, we may simply assume that $G = F/K$. Let $\phi : F \to F/K$ be the natural projection.

We may restrict ourselves to the case when K does not contain any lines. Indeed, let L be the maximal linear subspace of L contained in K (i.e. the union of all linear subspaces contained in K). We have $\bar{L} \subset K$ because K is closed. Since F is a locally convex vector group, it follows that $\bar{L}$ is a vector subspace of F, the proof being the same as for topological vector spaces. Thus $\bar{L} = L$. Let $\phi : F \to F/L$ be the canonical projection. The group F/L has a natural structure of a nuclear vector group; it is metrizable and complete. Evidently, $\psi(K)$ is a closed subgroup of F/L containing no lines. It remains to be observed that F/K is canonically topologically isomorphic to $(F/L)/\psi(K)$.

Let $U_1 \supset U_2 \supset \ldots$ be a fundamental sequence of neighbourhoods of zero in F, consisting of symmetric convex sets. In view of (9.3) and (2.14), we may assume that, for every n, the Minkowski functional

p_n of U_n is a pre-Hilbert seminorm on $M_n := \text{span } U_n$ with $B(p_n) = U_n$. We may also assume that

$$(1) \qquad \sum_{k=1}^{\infty} d_k^2(U_{n+1}, U_n \cap M_{n+1}) \leq 1$$

for every n. Denote $L_n = \text{span }(K \cap U_n)$ for $n = 1,2,\ldots$.

For every n, let Γ_n be the family of all linear functionals f on L_{n+1} satisfying the conditions

$$(2) \qquad f(K \cap U_{n+1}) \subset Z,$$

$$(3) \qquad \sup \{f(u) : u \in U_n \cap L_{n+1}\} < \infty.$$

We shall prove that Γ_n is at most countable. Denote $E = L_{n+1}$ and let r,s be the restrictions to L_{n+1} of p_n and p_{n+1}, respectively. We have $B_r = U_n \cap L_{n+1}$ and $B_s = U_{n+1} \cap L_{n+1}$. From (1) and (2.13) we obtain

$$(4) \qquad \sum_{k=1}^{\infty} d_k^2(B_s, B_r) \leq 1.$$

As usual, we have the canonical diagram

$$\begin{array}{ccc} E & \xrightarrow{\ \text{id}\ } & E \\ \downarrow \psi_s & & \downarrow \psi_r \\ E_s & \xrightarrow{\ \Lambda_{sr}\ } & E_r \end{array}$$

We shall prove that the mapping $h \to h\psi_r$ is a bijection of the set

$$\Omega = \{h \in E_r^* : h\psi_r(K \cap U_{n+1}) \subset Z\}$$

onto Γ_n.

If $h \in \Omega$, then it follows immediately from our definitions that $h\psi_r \in \Gamma_n$. So, take an arbitrary $f \in \Gamma_n$. We have $f(u) = 0$ for all $u \in \ker \psi_r$; this follows easily from (3). Consequently, there is some $h \in E_r^{\#}$ with $h\psi_r = f$. From (3) we see that h is bounded and (2) yields $h\psi_r(K \cap U_{n+1}) \subset Z$. Thus $h \in \Omega$. Finally, if $h_1 \neq h_2$, then

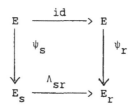

$h_1\psi_r \neq h_2\psi_r$ because ψ_r is surjective.

From (4) and (2.12) it follows that $d_k(\Lambda_{sr}) \to 0$ as $k \to \infty$, which means that Λ_{sr} is a compact operator. Therefore E_r is separable and, consequently, E_r^* is a separable Hilbert space. It is obvious that Ω is an additive subgroup of E_r^*. Furthermore, $\Omega \cap \text{int}\, B(E_r^*) = \{0\}$. Indeed, if $h \in \Omega \setminus \{0\}$, then $h(u) \neq 0$ for a certain $u \in \psi_r(K \cap U_{n+1})$ because $E = \text{span}\,(K \cap U_{n+1})$. But $h(u) \in Z$ and

$$\psi_r(K \cap U_{n+1}) \subset \psi_r(E \cap U_n) = B(U_r),$$

which implies that $\|h\| \geq 1$. Being a discrete subgroup of a separable space, Ω is at most countable. Consequently, so is Γ_n.

Let $\mathcal{B}$ be the family of all sets of the form

$$U_{n,f} = U_n + \{u \in L_{m+1} : f(u) = 0\}$$

where $m,n = 1,2,\ldots$ and $f \in \Gamma_m$. Observe that $\mathcal{B}$ is at most countable. It is evident that $\mathcal{B}$ satisfies conditions (a) - (c) of (1.12). So, there is a unique topology τ on F such that F_τ is a topological group for which $\mathcal{B}$ is a base at zero. Notice that if $U \in \mathcal{B}$, then $cU \in \mathcal{N}_o(F_\tau)$ for each $c > 0$. The original topology on F is finer than τ; we shall denote it by σ. Thus we may write $G = F_\sigma/K$.

We shall prove that F_τ is a metrizable nuclear vector group. To prove that F_τ is separated, take any $w \in F \setminus \{0\}$. We have to find some $U \in \mathcal{N}_o(F_\tau)$ with $w \notin U$. Suppose first that $w \notin K$. Since K is a closed subgroup of the nuclear vector group F_σ, from (9.4) and (8.6) it follows that there is some $\chi \in F_\sigma^\wedge$ with $\chi_{|K} \equiv 0$ and $\chi(w) \neq 0$. By (1.4), there is an index n such that $|\chi(U_n)| \leq \frac{1}{4}$. Next, according to (2.2), there is some $f \in M_n^\#$ with $\rho f = \chi_{M_n}$ and

$$(5) \qquad \sup\,\{|f(u)| : u \in U_n\} \leq \tfrac{1}{4}.$$

We have $f(K \cap M_n) \subset Z$ because $\chi_{|K} \equiv 0$. Therefore $f' := f_{|L_{n+1}} \in \Gamma_n$. Since $\rho f(w) = \chi(w) \neq 0$, it follows that $c := |f(w)| \neq 0$. By (5), for each $v \in U_n$, we have

$$|f(w - cv)| \geq |f(w)| - c|f(v)| \geq c - \tfrac{1}{4}c = \tfrac{3}{4}c > 0,$$

which means that $w \notin cU_n + f^{-1}(0)$. Hence $w \notin cU_{n,f'}$, either.

Now, suppose that $w \in K$. Since $w \neq 0$ and K does not contain any lines, there is some $t \in (0,1)$ with $tw \notin K$. According to the above, there is some radial $U \in \mathcal{N}_o(F_\tau)$ with $tw \notin U$. Hence $w \notin U$, either.

Being separated, F_τ is metrizable. Moreover, F is a locally convex vector group because all sets $U_{n,f}$ in $\mathcal{B}$ are symmetric and convex. Take any $m,n = 1,2,\ldots$ and any $f \in \Gamma_m$. Since F_σ is a nuclear vector group, there is an index $l > n$ such that $d_k(U_l,U_n) < k^{-1}$ for every k. Set $N = \{u \in L_{m+1} : f(u) = 0\}$. According to (2.6) (a), for every k, we have

$$d_k(U_{l,f},U_{n,f}) = d_k(U_l + N, U_n + N) \leq d_k(U_l, U_n + N) + d_l(N, U_n + N)$$

$$\leq d_k(U_l,U_n) + d_l(N,N) < k^{-1}$$

because $d_l(N,N) = 0$. This proves that F_τ is a nuclear vector group.

Our next goal is to show that the mapping $\phi : F_\tau \to F_\sigma/K$ is continuous. Take an arbitrary index n. We have to show that

(6) $\quad U_n + K \in \mathcal{N}_o(F)$.

Let r and s be defined as before. From (4) and (3.18) we have

$$L_{n+1} := \operatorname{span}(K \cap U_{n+1}) = \operatorname{span}(K \cap B_s) \subset \tfrac{1}{2}B_r + \operatorname{gp}(K \cap B_s)$$

$$\subset \tfrac{1}{2}U_n + K.$$

Hence $\frac{1}{2}U_n + L_{n+1} \subset U_n + K$, which proves (6) because, evidently, $\frac{1}{2}U_n + L_{n+1} \in \mathcal{N}_o(F_\tau)$.

The completion $\tilde{F}_\tau$ of F_τ is a nuclear vector group due to (9.5). Let $\tilde{\phi} : \tilde{F}_\tau \to G$ be the canonical extension of ϕ.

Without loss of generality we may assume that $\Sigma g_i = 0$; then $P = \mathcal{S}(\Sigma g_i; G)$. Since $G = F_\sigma/K$, we can find a null sequence $(s_i)_{i=1}^\infty$ in F_σ with $\phi(s_j) = \sum_{i=1}^{j} g_i$ for $j = 1,2,\ldots$. Set $u_1 = s_1$ and $u_i = s_i - s_{i-1}$ for $i = 2,3,\ldots$. Then

(7) $\quad \sum_{i=1}^{j} u_i = s_j \xrightarrow[j\to\infty]{} 0 \quad$ in $\quad F_\sigma$

and $\phi(u_i) = g_i$ for $i = 1,2,\ldots$. From (10.8) it follows that $Q := \mathcal{S}(\Sigma u_i;\tilde{F}_\tau)$ is a closed linear subspace of the nuclear Fréchet space $(\tilde{F}_\tau)_o$.

It remains to show that $\tilde{\phi}(Q) = P$. The inclusion $\tilde{\phi}(Q) \subset P$ is trivial. To prove the opposite one, choose any $a \in P$. There is a permutation π of positive integers such that the series $\Sigma g_{\pi(i)}$ converges to a. Choose some $w \in F$ with $\phi(w) = a$. We have

$$\phi(w - \sum_{i=1}^{j} u_{\pi(i)}) = a - \sum_{i=1}^{j} g_{\pi(i)} \xrightarrow[j\to\infty]{} 0.$$

Consequently, there is a sequence $(z_j)_{j=1}^{\infty}$ in K such that

$$z_j + \sum_{i=1}^{j} u_{\pi(i)} \xrightarrow[j\to\infty]{} w \quad \text{in} \quad F_\sigma.$$

Define $v_1 = z_1$ and $v_i = z_i - z_{i-1}$ for $i = 2,3,\ldots$. Then $v_i \in K$ for every i, and

(8) $$\sum_{i=1}^{j} [u_{\pi(i)} + v_i] \xrightarrow[j\to\infty]{} w \quad \text{in} \quad F_\sigma.$$

From (7) and (8) it follows that $(u_i)_{i=1}^{\infty}$ and $(u_{\pi(i)} + v_i)_{i=1}^{\infty}$ are null sequences in F_σ. Hence $(v_i)_{i=1}^{\infty}$ is a null sequence in F_σ, too.

We shall prove that the series Σv_i satisfies Cauchy's criterion in F_τ. Take any $m,n = 1,2,\ldots$ and any $f \in \Gamma_m$. There is an index j_1 such that $v_i \in U_{m+1}$ for $i > j_1$; then $v_i \in L_{m+1}$ because $v_i \in K$. Consequently, we have $f(v_i) \in Z$ for $i > j_1$. Since $f \in \Gamma_m$ and $v_i \to 0$ in F_σ, replacing n by m in (3), we see that $f(v_i) \to 0$. Hence there is an index $j_2 > j_1$ such that $f(v_i) = 0$ for $i > j_2$. This means that $\sum_{i=j_2}^{j} v_i \in U_{n,f}$ for each $j > j_2$.

Since $\tilde{F}_\tau$ is complete, the series Σv_i converges in $\tilde{F}_\tau$ to a certain point y. Let $\bar{K}$ be the closure of K in $\tilde{F}_\tau$; then $y \in \bar{K}$ because all v_i were in K. Consequently,

$$\tilde{\phi}(y) \in \tilde{\phi}(\bar{K}) \subset \overline{\tilde{\phi}(K)} = \overline{\phi(K)} = \overline{\{0\}} = \{0\}.$$

From (8) we infer that the series $\Sigma u_{\pi(i)}$ converges to $w - y$ in $\tilde{F}_\tau$. Thus $w - y \in Q$. Finally,

$$\tilde{\phi}(w - y) = \tilde{\phi}(w) - \tilde{\phi}(y) = \phi(w) - 0 = a,$$

which proves that $P \subset \tilde{\phi}(Q)$. ∎

(10.10) EXAMPLE. Let $\psi : R^2 \to T^2$ be the canonical projection. Take some $u = (x,y) \in R^2$ with y/x irrational and denote $L = \text{span}\,\{u\}$. Next, set $u_i = \frac{(-1)^i}{i}u$ and $g_i = \psi(u_i)$ for every i. It is not hard to see that $\mathfrak{S}(\Sigma u_i;R^2) = L$ and $P := \mathfrak{S}(\Sigma g_i;T^2) = \psi(L)$. Thus P is a dense, but non-closed subgroup of T^2. Notice that $\mathfrak{C}(\Sigma g_i) = \mathfrak{A}(\Sigma g_i) = T^2$.

(10.11) REMARKS. Let Σu_i be a convergent series in a real topological vector space E. Let us denote

$$\Gamma(\Sigma u_i) = \{f \in E^* : \sum_{i=1}^{\infty} |f(u_i)| < \infty\}.$$

Then the set

$$\Gamma_0(\Sigma u_i) = \{u \in E : f(u) = 0 \ \text{ for all } \ f \in \Gamma(\Sigma u_i)\}$$

is a closed linear subspace of E.

The Lévy-Steinitz theorem is often formulated in the following, somewhat stronger version: for each convergent series Σu_i in R^n, one has

$$(1) \qquad \mathfrak{S}(\Sigma u_i) = \Sigma u_i + \Gamma_0(\Sigma u_i).$$

This result was obtained by E. Steinitz [87]; it is sometimes called the Steinitz theorem. However, in the literature there is no consequence, and the expressions "Steinitz theorem" and "Lévy-Steinitz theorem" are used exchangeably. It was proved in [10] that (1) holds for each convergent series Σu_i in a metrizable nuclear space. Example (10.10) shows that there is no reasonable way of extending this result to series in nuclear groups.

It has been proved in [14] that if a metrizable locally convex space is not nuclear, then it contains a convergent series Σu_i such that $\mathfrak{A}(\Sigma u_i)$ is not a linear subspace. Then (1) does not hold, for, in view of (10.4) and the obvious inclusion $\mathfrak{A}(\Sigma u_i) \subset \Gamma_0(\Sigma u_i)$, we have

$$\mathfrak{S}(\Sigma u_i) \subset \mathfrak{C}(\Sigma u_i) = \Sigma u_i + \mathfrak{A}(\Sigma u_i) \subset \Sigma u_i + \Gamma_0(\Sigma u_i).$$

It is not known whether the following sentence is true: if a metrizable locally convex space is not nuclear, then it contains a convergent series Σu_i such that $\mathcal{S}(\Sigma u_i)$ is not a linear manifold.

(10.12) REMARKS. In infinite dimensional Banach spaces, the Lévy--Steinitz theorem fails to hold (Problem 106 from the Scottish Book; see [81], p. 188). The solution of this problem has an interesting history; see [34] or [43], pp. 44-45. V.M. Kadets [45] proved that every infinite dimensional Banach space contains a convergent series with a non-convex set of sums. See also [46], especially Theorem 10.

Some years ago M.I. Kadets found an interesting example of a series in a Hilbert space, making the conjecture that the set of sums of this series consists of exactly two points. The conjecture was proved independently by K. Woźniakowski and P.A. Kornilov; see [44]. Making use of this example and applying a standard technique, one can construct a series with n-point set of sums in every infinite dimensional normed space; see [46], Theorem 10. Another example of a series with two-point set of sums was obtained by P. Enflo.

(10.13) REMARK. There are several analogous of the Lévy-Steinitz theorem which are valid in infinite dimensional Banach spaces. They assert that if Σu_i is a convergent series in a Banach space E, then condition (1) of (10.11) is satisfied under various additional assumptions on E and Σu_i. A typical example is the situation when E is a Hilbert space and $\Sigma \|u_i\|^2 < \infty$. The best source of information here is [46], pp. 158-159.

(10.14) REMARK. Let $R^{(0,1)}$ be the space of all real-valued functions on the unit interval, endowed with the topology of pointwise convergence (i.e. the product of continuum real lines). The example of M.I. Kadets mentioned in (10.12) allows one to construct a series in $R^{(0,1)}$ with two-point set of sums (see [43], Theorem 6.4.3, p. 172). Applying (10.9), we see that the product $T^{(0,1)}$ of continuum circles contains a convergent series such that its set of sums consists of exactly two points and is not a coset modulo any subgroup of $T^{(0,1)}$ (cf. (10.2)).

(10.15) REMARK. From the results of this section it follows easily that if (g_i) is an arbitrary null sequence in a metrizable and complete nuclear group, then there exist a permutation π of indices and a sequence of signs $\varepsilon_i = \pm 1$ such that the series $\Sigma \varepsilon_i g_{\pi(i)}$ is convergent. On the other hand, it can be shown that if a metrizable locally convex space (and probably even a locally quasi-convex group) is not nuclear, then it contains a null sequence (u_i) such that the series $\Sigma \varepsilon_i u_{\pi(i)}$ is divergent for each permutation π and each sequence $\varepsilon_i = \pm 1$.

(10.16) REMARK. Let $(g_i)_{i \in I}$ be a system of elements of an abelian topological group G. We say that $(g_i)_{i \in I}$ satisfies the Cauchy criterion of unconditional convergence if to each $U \in N_o(G)$ there corresponds a finite subset J of I such that $\sum_{i \in K} g_i \in U$ for each subset K of $I \setminus J$. Next, we say that the system $(g_i)_{i \in I}$ is absolutely summable if $\sum_{i \in I} (g_i/U) < \infty$ for each $U \in N_o(G)$. It turns out that if a system of elements of a nuclear group satisfies the Cauchy criterion of unconditional convergence, then it is absolutely summable. Hence, every unconditionly convergent series in a complete nuclear group is absolutely convergent. The proof will be given in a separate paper.

(10.17) REMARK. By the weak topology on an abelian topological group G we mean the topology induced by the family of all continuous characters of G. If G is a locally convex space, this topology is much weaker than the weak topology induced by the family of all continuous linear functionals, but defines the same class of convergent sequences (it is enough to consider the case $G = R$).

A series Σg_i in G is said to be <u>subseries convergent</u> if the series $\Sigma \varepsilon_i g_i$ is convergent for each sequence $\varepsilon_i = 0,1$. The Orlicz--Pettis theorem says that if a series in a locally convex space is subseries convergent in the weak topology, then it is subseries convergent in the original topology, too. The same is true for series in a separable locally quasi-convex group; this follows directly from Theorem 7 of [48]. Hence, in view of (8.6), the Orlicz-Pettis theorem remains valid for series in arbitrary nuclear groups.

It seems very likely that every weakly convergent sequence in a nuclear group is convergent in the original tcpology.

(10.18) REMARKS. Let $\Phi : E \to F$ be a linear operator acting between normed spaces. By $\beta(\Phi)$ we shall denote the smallest number

$r > 0$ with the following property:

to each finite system $u_1,\dots,u_n \in B_E$ there corresponds signs $\varepsilon_1,\dots,\varepsilon_n = \pm 1$ such that $\|\Phi(\varepsilon_1 u_1 + \dots + \varepsilon_n u_n)\| \leq r$.

We say that Φ is a <u>balancing operator</u> if $\beta(\Phi) < \infty$.

The proof of (10.7) is based on (7.13). Lemma (7.13) also occurs in the proof of (10.5). It is not hard to see that (7.13) says, in fact, that Hilbert-Schmidt operators are balancing. More precisely, if Φ acts between unitary spaces, then $\beta(\Phi) = [\sum_{k=1}^{\infty} d_k^2(\Phi)]^{1/2}$; the proof very similar to that of (7.13), is given in [13]. On the other hand, it can be shown that if Φ is an arbitrary operator acting between normed spaces, then $\beta(\Phi) \geq C \sup_k k^{1/2} v_k(\Phi)$ where C is a universal constant (the numbers $v_k(\Phi)$ were defined in section 6). It is a standard fact that every finite dimensional operator Φ is balancing, with $\beta(\Phi) \leq 2\|\Phi\| \operatorname{rank} \Phi$. The Beck-Fiala theorem [20] implies that $\beta(\mathrm{id} : l^1 \to l^\infty) \leq 2$. Beck and Spencer [21] proved that the diagonal operator $\Phi : l^\infty \to l^\infty$ given by $\Phi e_k = k^{-1/2}(\log k)^{-1} e_k$ is balancing. Spencer [86] showed that the factor $(\log k)^{-1}$ can be omitted. It is an open problem whether the canonical embedding of l^2 into l^∞ is balancing (this is called the Komlós conjecture).

By $\phi(\Phi)$ we denote the smallest number $r > 0$ with the following property:

to each finite system $u_1,\dots,u_n \in B_E$ with $u_1 + \dots + u_n = 0$ there corresponds a permutation π of indices, such that

$$\|\sum_{i=1}^{j} \Phi u_{\pi(i)}\| \leq r \qquad (j = 1,\dots,n).$$

We say that Φ is a <u>Steinitz operator</u> if $\phi(\Phi) < \infty$.

Every finite dimensional operator Φ is a Steinitz operator with $\phi(\Phi) \leq \|\Phi\| \operatorname{rank} \Phi$ (see [34], [32] and [10], Remark 3). Bárány [16] proved that the diagonal operator $\Phi : l^\infty \to l^\infty$ given by $\Phi e_k = 2^{-3^k} e_k$ is a Steinitz operator with $\phi(\Phi) \leq 1$. Lemma (10.5) says that if Φ acts between unitary spaces, then $\phi(\Phi) \leq \sum_{k=1}^{\infty} d_k(\Phi)$. It is not hard to see that $\beta(\Phi) \leq 2\phi(\Phi)$ for every Φ.

By $\sigma(\Phi)$ we denote the smallest number $r > 0$ with the following property:

to each finite sequence $u_1,\ldots,u_n \in B_E$ there corresponds a sequence of signs $\varepsilon_1,\ldots,\varepsilon_n = \pm 1$ such that

$$\|\sum_{i=1}^{j} \varepsilon_i \Phi u_i\| \leq r \qquad (j = 1,\ldots,n).$$

We say that Φ is a <u>strongly balancing operator</u> if $\sigma(\Phi) < \infty$.

Every finite dimensional operator Φ is strongly balancing with $\sigma(\Phi) \leq 2\|\Phi\|$ rank Φ (see [17]; in fact, $\sigma(\Phi) \leq \|\Phi\|(2 \text{ rank } \Phi - 1)$). J. Beck [19] proved that the diagonal operator $\Phi : l^\infty \to l^\infty$ given by $\Phi e_k = k^{-(2+\varepsilon)\log k} e_k$ is strongly balancing for each fixed $\varepsilon > 0$. It seems likely that the exponent $(2 + \varepsilon) \log k$ can be replaced here by some positive constant independent of k. The existence of such a constant would imply that, for each null sequence (u_i) in a nuclear Fréchet space, one can choose signs $\varepsilon_i = \pm 1$ such that the series $\Sigma\varepsilon_i u_i$ is convergent (for countable products of real lines this fact was proved by Katznelson and McGehee [51]). What is more, a nuclear Fréchet space could be replaced here by a metrizable and complete nuclear group. On the other hand, if a metrizable locally convex space is not nuclear, then it contains a null sequence (u_i) such that the series $\Sigma\varepsilon_i u_i$ is divergent for each sequence $\varepsilon_i = \pm 1$ (cf. (10.15)).

By definition, one has $\beta(\Phi) \leq \sigma(\Phi)$. Pecherskiĭ [74], Lemma 1, proved that, up to the notation, $\phi(\Phi) \leq 3\sigma(\Phi)$. A very short and simple proof of the inequality $\phi(\Phi) \leq \sigma(\Phi)$ was found by S. Chobanyan (unpublished). Thus every strongly balancing operator is a Steinitz operator.

(10.19) NOTE. The results of this section are new. The method of the proof of (10.8) is taken from [10]. The argument applied to obtain (10.8) from (10.5) through (10.6) and (10.7) is standard. In various forms, it occurs in several papers on infinite dimensional generalizations of the Lévy-Steinitz theorem; see (10.13). Its main idea goes back to Steinitz [87]. Lemma (10.7) is a straightforward consequence of the results of [20] or [21]; cf. [10], Remark 2.

Chapter 4

THE BOCHNER THEOREM

In this chapter we show that nuclear groups satisfy Bochner's theorem on positive-definite functions. Section 11 wears an introductory complexion. We introduce here some new terminology and state several standard results in a form convenient to us. Section 12 contains the proof of the main result. Finally, in section 13 we give some applications. We formulate here an appropriate version of the SNAG theorem for nuclear groups and prove that each continuous positive-definite function (resp. continuous unitary representation) defined on a subgroup of a nuclear group can be extended to the whole group.

11. Preliminaries

Bochner's classical theorem asserts that each continuous positive-definite function on the real line is the Fourier transform of some Radon measure. Many far-reaching generalizations of this fact are known. Roughly speaking, they say that, under certain assumptions on a topological group G, each continuous positive-definite function on G may be written as an integral of some measure on the dual object $G^{\wedge}$ (cf. [63], Ch. VI, §9).

We confine ourselves to abelian groups only. By a Bochner theorem for an abelian group G we shall mean a statement of the following form:

(*) each continuous positive-definite function on G is the Fourier transform of a (unique) Radon measure on $G^{\wedge}$.

There are at least three situations where (*) is known to be true:

(i) the Weil-Raikov theorem asserts that (*) holds for any LCA group G;

(ii) the fact that (*) is true for every nuclear locally convex space G is known as the Minlos theorem (see [67], Theorem 1, p. 508 or [63], Ch. IV, Theorem 4.3, p. 318); it follows easily from the Minlos theorem that (*) is true when G is a Hausdorff quotient group of a nuclear locally convex space (see Yang [104]);

(iii) each locally convex space G over p-adic field satisfies (*); this was proved by Madrecki [62].

We do not specify here the topology on $G^\wedge$. The Weil-Raikov theorem is usually formulated for the compact-open topology, whereas the Minlos theorem in the language of vector spaces, the topology on the dual space G^* being the topology of uniform convergence on finite, bounded or, say, compact convex subsets of G (there is a canonical isomorphism between G^* and $G^\wedge$; see (2.3)). The result of (iii) was proved for the weak* topology on G^*.

We shall prove that (*) holds for every nuclear group G (Theorem (12.1)); this is a common generalization of (i) - (iii).

The Bochner theorm in the form of (*) characterizes nuclear spaces among metrizable locally convex spaces ([70], Theorem 5, p. 75). On the other hand, there is version of the Bochner theorem which is valid in any locally convex space ([30], Theorem 1, p. 348). Thus, each continuous positive-definite function on a locally convex space can be in some way synthesized of continuous characters. The situation becomes completely different when we start to consider quotient groups. In section 5 we gave an example of a discrete subgroup K of the space l^p, $p > 2$, such that the quotient group l^p/K admits non-trivial continuous unitary representations but does not admit any non-trivial continuous characters. Thus, there are on l^p/K continuous positive-definite functions which cannot be synthesized of continuous characters (since the latter do not exist); therefore one cannot speak of any version of Bochner's theorem in this case. It is quite possible that similar examples can be constructed in any non-nuclear locally convex metrizable space.

Let G be an abelian topological group. We say that τ is an <u>admissible topology</u> on $G^\wedge$ if the mappings $G^\wedge_\tau \ni \chi \to \chi(g)$, $g \in G$, are continuous and the sets U^o_G, $U \in N_o(G)$, are compact in $G^\wedge_\tau$. From (1.5) it follows that the topologies of pointwise, compact and precompact convergence are all admissible.

Let X be a topological space. The family of Borel subsets of X is denoted by $\mathcal{B}(X)$. By a <u>Borel measure</u> on X we mean a σ-additive mapping of $\mathcal{B}(X)$ into $[0,\infty]$. A finite Borel measure μ on X is called a <u>Radon measure</u> if, for each $A \in \mathcal{B}(X)$ and each $\varepsilon > 0$, there exists a compact subset Q of A with $\mu(A \setminus Q) < \varepsilon$.

Let X,Y be two topological spaces, $\pi : X \to Y$ a Borel mapping and μ a Borel measure on X. Then the mapping $\mathcal{B}(Y) \ni A \to \mu(\pi^{-1}(A))$ is a Borel measure on Y. We call it the π-image of μ and denote by μ_π. If f is a μ_π-integrable function on Y, then $f\pi$ is a μ-integrable function on X and

$$\int_X f\pi d\mu = \int_Y f d\mu_\pi .$$

Let G be an abelian topological group and let τ be a topology on $G^{\wedge}$ such that all the mappings $G^{\wedge}_{\tau} \ni \chi \to \chi(g)$, $g \in G$, are continuous. By the <u>Fourier transform</u> of a finite Borel measure μ on $G^{\wedge}_{\tau}$ we mean the function

$$G \ni g \to \int_{G^{\wedge}_{\tau}} \exp\,[2\pi i\chi(g)]d\mu(\chi).$$

The Fourier transform of μ is denoted by $\hat{\mu}$; it is sometimes called the inverse Fourier-Stieltjes transform of μ. A Borel measure μ on $G^{\wedge}_{\tau}$ will be called <u>regular</u> if, for each $A \in \mathcal{B}(G^{\wedge}_{\tau})$ and each $\varepsilon > 0$, there exists a compact equicontinuous subset Q of A with $\mu(A \setminus Q) < \varepsilon$.

(11.1) PROPOSITION. Let G be an abelian topological group and let τ be a topology on $G^{\wedge}$ such that the mappings $G^{\wedge}_{\tau} \ni \chi \to \chi(g)$, $g \in G$, are continuous. If μ is a finite Borel measure on $G^{\wedge}_{\tau}$, then $\hat{\mu}$ is a p.d. function on G with $\hat{\mu}(0) = \mu(G^{\wedge}_{\tau})$. If μ is regular, $\hat{\mu}$ is continuous.

Proof. The fact that $\hat{\mu}$ is a p.d. function is well known (see e.g. [38], (33.1)). So, assume that μ is regular and take any $\varepsilon > 0$. There is an equicontinuous subset Q of $G^{\wedge}_{\tau}$ with $\mu(\setminus Q) < \varepsilon$. Next, we can find some $U \in N_o(G)$ such that $|\chi(U)| < \varepsilon$ for all $\chi \in Q$. Then, for each $g \in U$, we have

$$\begin{aligned} |\hat{\mu}(g) - \hat{\mu}(0)| &= \Big|\int_{G^{\wedge}} (1 - \exp\,[2\pi i\chi(g)])d\mu(\chi)\Big| \\ &\le \int_{G^{\wedge}} |1 - \exp\,[2\pi i\chi(g)]|d\mu(\chi) \le \int_{G^{\wedge}} 2\pi|\chi(g)|d\mu(\chi) \\ &= 2\pi \int_{Q} |\chi(g)|d\mu(\chi) + 2\pi \int_{\setminus Q} |\chi(g)|d\mu(\chi) \\ &\le 2\pi\varepsilon\mu(Q) + \pi\mu(\setminus Q) < [2\mu(G^{\wedge}) + 1]\pi\varepsilon. \end{aligned}$$

Thus $\hat{\mu}$ is continuous at zero and the continuity at the remaining points follows from (1.22) (c). ∎

(11.2) PROPOSITION. Let $\phi : G \to H$ be a continuous homomorphism of abelian topological groups. Suppose that the dual groups $G^{\wedge}$ and $H^{\wedge}$ are endowed with some topologies such that the mappings $G^{\wedge} \ni \chi \to \chi(g)$, $g \in G$, and $H^{\wedge} \ni \kappa \to \kappa(h)$, $h \in H$, are continuous. Suppose also that the dual homomorphism $\psi = \phi^{\wedge} : H^{\wedge} \to G^{\wedge}$ is continuous. If μ is a finite Borel measure on $H^{\wedge}$, then $\hat{\mu}_{\psi} = \hat{\mu}\phi$.

Proof. For each $g \in G$, one has

$$\hat{\mu}_\psi(g) = \int_{G^\wedge} \exp[2\pi i\langle\chi,g\rangle]\,d\mu_\psi(\chi) = \int_{H^\wedge} \exp[2\pi i\langle\psi(\chi),g\rangle]\,d\mu(\kappa)$$
$$= \int_{H^\wedge} \exp[2\pi i\langle\kappa,\phi(g)\rangle]\,d\mu(\kappa) = \hat{\mu}(\phi(g)). \quad \blacksquare$$

(11.3) PROPOSITION. Let G be an abelian topological group and let τ be a topology on $G^\wedge$ such that the mappings $G^\wedge_\tau \ni \chi \to \chi(g)$, $g \in G$, are continuous. If μ_1, μ_2 are two Radon measures on $G^\wedge_\tau$ with $\hat{\mu}_1 = \hat{\mu}_2$, then $\mu_1 = \mu_2$.

Proof. Let G_d denote the group G endowed with the discrete topology. The identity homomorphism $\phi : G^\wedge \to (G_d)^\wedge_c$ is continuous. Let ν_i be the ϕ-image of μ_i, $i = 1,2$. Then ν_i is a Radon measure on $(G_d)^\wedge_c$ and (11.2) implies that $\hat{\nu}_i = \hat{\mu}_i$. Thus $\hat{\nu}_1 = \hat{\nu}_2$. From the uniqueness of the measure in the Weil-Raikov theorem it now follows that $\nu_1 = \nu_2$. If Q is a compact subset of $G^\wedge_\tau$, then $\phi(Q)$ is a compact, hence Borel, subset of $(G_d)^\wedge_c$, and $\mu_i(Q) = \mu_1(\phi^{-1}(\phi(Q))) = \nu_i(\phi(Q))$ for $i = 1,2,$. Hence $\mu_1(Q) = \mu_2(Q)$. This completes the proof because μ_1, μ_2 are both Radon measures. ∎

(11.4) NOTE. The material of this section is standard.

12. The Bochner theorem

The aim of this section is to prove the following fact:

(12.1) THEOREM. Let G be a nuclear group and τ an admissible topology on $G^\wedge$. Then the mapping $\mu \to \hat{\mu}$ establishes a one-to-one correspondence between the family of all regular finite Borel measures on $G^\wedge_\tau$ and the family of all continuous p.d. functions on G.

Let E be a nuclear space and E^* the dual space endowed with the topology of uniform convergence on finite, compact or precompact sets. It is not difficult to see that the topology induced on $E^\wedge$ by the canonical homomorphism $\rho_E : E^* \to E^\wedge$ is an admissible one. Thus (12.1) implies the Minlos theorem.

The proof given below is patterned upon that of the Minlos theorem. The main difference lies in replacing the Minlos lemma ([67], Lemma 4, p. 510) by its analogue for p.d. functions on additive subgroups of R^n (lemma (12.2) below). To obtain (12.1) from (12.2) is a matter of technique; we have found it most convenient to aply here the Prokhorov theorem on inverse limits of measures, in the form due to Kisyński [54].

(12.2) **LEMMA.** Let N be an n-dimensional vector space, $n \geq 1$, and let p,q be two pre-Hilbert seminorms on N such that

$$\sum_{k=1}^{n} d_k^{1/2}(B_p,B_q) < \frac{1}{12}.$$

Let K be a subgroup of N and μ a Borel measure on $K_p^{\frown}$ with $\mu(K^{\frown}) = 1$. Suppose that there is some $\varepsilon > 0$ such that $\operatorname{Re}\hat{\mu}(u) \geq 1-\varepsilon$ for all $u \in K \cap B_q$. Then $\mu((K \cap B_p)_K^{\circ}) \geq 1 - 2\varepsilon$.

Proof. Suppose first that both p and q are norms. In this case, we may simply assume that $N = R^n$ and q is the euclidean norm on N. Then $B_q = B_n$ and B_p is some ellipsoid in R^n; let $\xi_n,\ldots,\xi_n$ be its principal semiaxes. Obviously, we have

$$(1) \qquad \sum_{k=1}^{n} \xi_k^{1/2} = \sum_{k=1}^{n} d_k^{1/2}(B_p,B_q) < \frac{1}{12}.$$

We begin with the case when K is discrete. Let $L = \operatorname{gp}(K \cap B_p)$ and let $\pi : K_p^{\frown} \to L_p^{\frown}$ be the natural homomorphism. Denote $\nu = \mu_\pi$. According to (11.2), we have $\hat{\nu} = \hat{\mu}_{|L}$. We shall prove that

$$(2) \qquad \nu((L \cap B_p)_L^{\circ}) \geq 1 - 2\varepsilon.$$

Let $M = \operatorname{span} L$ and $m = \dim M$. If $L = \{0\}$, then

$$\nu((L \cap B_p)_L^{\circ}) = \nu(\{0\}_L^{\circ}) = \nu(L^{\frown}) = \hat{\nu}(0) = \hat{\mu}(0) \geq 1 - \varepsilon > 1 - 2\varepsilon.$$

So, suppose that $m \geq 1$. Let $\eta_1,\ldots,\eta_m$ be the principal semiaxes of the ellipsoid $D = M \cap B_p$. From (1) and (2.13) it follows that

$$(3) \qquad \sum_{k=1}^{m} \eta_k^{1/2} < \frac{1}{12}.$$

We may assume that $M = R^n$ and

$$D = \{(x_1,\ldots,x_m) \in R^m : \sum_{k=1}^{m} (x_k/\eta_k)^2 \leq 1\}.$$

Set $\zeta_k = \eta_k^{3/4}$ for $k = 1,\ldots,m$ and denote

$$E = \{(x_1,\ldots,x_m) \in R^m : \sum_{k=1}^{m} (x_k/\zeta_k)^2 \leq 1\}.$$

We have $L = \mathrm{gp}\,(L \cap D)$ and, by (3),

$$\sum_{k=1}^{m} d_k^2(D,E) = \sum_{k=1}^{m} (\eta_k/\zeta_k)^2 = \sum_{k=1}^{m} \eta_k^2 < \frac{1}{12} \leq 4.$$

According to (3.17), we can find a rectangular parallelotope $P \subset E$ such that $\{u+P\}_{u\in L}$ is a disjoint covering of R^m. Set $a_k = \eta_k^{1/4}$ for $k = 1,\ldots,m$ and denote

$$A = \{(x_1,\ldots,x_m) \in R^m : -a_k < x_k < a_k \text{ for } k = 1,\ldots,m\}.$$

From (3) we get $a_1^2 + \ldots + a_m^2 < 1/12$, which implies that $A \subset B_q$. Therefore, for each $u \in L \cap A$, we have $\mathrm{Re}\,\hat{\nu}(u) = \mathrm{Re}\,\hat{\mu}(u) \geq 1 - \varepsilon$, i.e.

(4) $$\mathrm{Re} \int_{L^\frown} \exp\,[2\pi i\kappa(u)]d\nu(\kappa) \geq 1 - \varepsilon.$$

let us write $|W| = \mathrm{card}\,(L \cap W)$ for $W \subset M$. Consider the function

$$L^\frown \ni \kappa \to f(\kappa) = \frac{1}{|A|} \sum_{u\in L\cap A} \mathrm{Re}\,\exp\,[2\pi i\kappa(u)]$$

and denote $X = \{\kappa \in L^\frown : f(\kappa) \geq \frac{1}{2}\}$. From (4) we obtain

$$\int_{L^\frown} f(\kappa)d\nu(\kappa) \geq 1 - \varepsilon,$$

whence $\nu(X) \geq 1 - 2\varepsilon$ because $f \leq 1$. Thus, to prove (2), it remains to show that $X \subset (L \cap B_p)^{\circ}_{L}$.

To this end, take any $\kappa \in X$ and $w = (w_1,\ldots,w_m) \in L \cap D$. We are to show that

(5) $$|\kappa(w)| \leq \frac{1}{4}.$$

Since $P \subset E$, and the sets $u + P$, $u \in L$, are pairwise disjoint, we have

$$|A|\ \mathrm{vol}_m(P) = \mathrm{vol}_m\,(\bigcup_{u\in L\cap A} (u + P)) \leq \mathrm{vol}_m\,(A + E).$$

It is clear that

$$A + E \subset \{(x_1,\ldots,x_m) \in R^m : -a_k - \zeta_k < x_k < a_k + \zeta_k \text{ for } k = 1,\ldots,m\},$$

whence

(6) $$|A| \operatorname{vol}_m (P) \le 2^m \sum_{k=1}^{m} (a + \zeta_k).$$

Let $A_w = w + A$ and $B = A \cap A_w$. The set $U = M \setminus ((M \setminus B) + E)$ is an m-dimensional rectangular parallelotope with edges equal resp. to $2a_k - 2\zeta_k - |w_k|$, $k = 1,\ldots,m$. Since $P \subset E$, and $\{u + P\}_{u\in L}$ is a covering of R^m, it follows that $U \subset \bigcup_{u\in L\cap B} (u + P)$. Hence

(7) $$|B| \operatorname{vol}_m (P) \ge \operatorname{vol}_m (U) = \prod_{k=1}^{m} (2a_k - 2\zeta_k - |w_k|).$$

Let V be the symmetric difference of A and A_w, i.e.

$$V = (A \setminus A_w) \cup (A_w \setminus A) = (A \setminus B) \cup (A_w \setminus B).$$

Let us write

$$s = \frac{1}{|A|} \sum_{u\in L\cap A_w} \exp [2\pi i\kappa(u)], \quad t = \frac{1}{|A|} \sum_{u\in L\cap A} \exp [2\pi i\kappa(u)].$$

Then $s = t \exp [2\pi i\kappa(w)]$. On the other hand,

$$|s - t| \le \frac{1}{|A|} \sum_{u\in L\cap V} |\exp [2\pi i\kappa(u)]| \le \frac{|V|}{|A|}.$$

It is clear that $|V| = 2(|A| - |B|)$, whence, by (6) and (7),

$$|s - t| \le 2[1 - \frac{|B|}{|A|}] \le 2[1 - \frac{\sum_{k=1}^{m} (2a_k - 2\zeta_k - |w_k|)}{2^m \prod_{k=1}^{m} (a_k + \zeta_k)}].$$

Hence, applying the standard inequalities

$$\frac{1 - x - y}{1 + x} \ge 1 - 2x - y \quad \text{and} \quad \prod_{k=1}^{m} (1 - x_k) \ge 1 - \sum_{k=1}^{m} x_k,$$

we obtain

$$|s - t| \le 4 \sum_{k=1}^{m} (\zeta_k/a_k) + 2 \sum_{k=1}^{m} (|w_k|/a_k).$$

Since $w \in D$, we have $\sum_{k=1}^{m} (w_k/\eta_k)^2 \le 1$, and thus, $|w_k| \le \eta_k$ for $k = 1,\ldots,m$. Therefore, by (1),

$$|s - t| \leq 4 \sum_{k=1}^{m} \eta_k^{1/2} + 2 \sum_{k=1}^{m} \eta_k^{3/4} \leq 6 \sum_{k=1}^{m} \eta_k^{1/2} < \frac{1}{2}.$$

On the other hand, we have

$$|s - t| = |t \exp [2\pi i \kappa(w)] - t| = |t| \cdot |1 - \exp [2\pi i \kappa(w)]|.$$

But $\kappa \in X$, therefore $\operatorname{Re} t = f(\kappa) \geq 1/2$ and $|t| \geq 1/2$. Hence we obtain (5) and, in consequence, (2). It is clear that $\pi^{-1}((L \cap B_p)_L^o) = (K \cap B_p)_K^o$. Then, by (2), we get

$$\mu((K \cap B_p)_K^o = \mu(\pi^{-1}((L \cap B_p)_L^o) = \nu((L \cap B_p)_L^o) \geq 1 - 2\varepsilon.$$

This completes the proof in the case when K is discrete.

Now, let K be an arbitrary subgroup of N. We may clearly assume K to be closed. Then K is the direct sum of some discrete subgroup and some linear subspace (see (3.1)) and, therefore, we can find an increasing sequence (K_m) of discrete subgroups of K such that their union is dense in K. For every m, let $\pi_m : K^\wedge \to K_m^\wedge$ be the natural homomorphism and let $\mu_m = \mu_{\pi_m}$. From (11.2) we get $\hat{\mu}_m = \hat{\mu}_{|K_m}$. Since, as we have just proved, the lemma is true for discrete subgroups, it follows that $\mu_m((K_m \cap B_p)^o) \geq 1 - 2\varepsilon$ for every m. The subgroups K_m may be chosen in such a way that the condition

$$\bigcap_{m=1}^{\infty} \pi_m^{-1}((K_m \cap B_p)^o) \subset (K \cap B_p)^o$$

be satisfied. Then

$$\begin{aligned}\mu((K \cap B_p)^o) &\geq \mu(\bigcap_{m=1}^{\infty} \pi_m^{-1}((K_m \cap B_p)^o)) \\ &= \lim_{m\to\infty} \mu(\pi_m^{-1}((K_m \cap B_p)^o)) = \lim_{m\to\infty} \mu_m((K_m \cap B_p)^o) \\ &\geq 1 - 2\varepsilon.\end{aligned}$$

This completes the proof in the case when p and q are norms.

Finally, consider the case when p and q are arbitrary pre--Hilbert seminorms on N. Then, clearly, one can find two decreasing sequences (p_m) and (q_m) of unitary norms on N such that

$$\sum_{k=1}^{n} d_k^{1/2} (B_{p_m}, B_{q_m}) < \frac{1}{12} \qquad (m = 1,2,\ldots),$$

$B_{q_m} \subset B_q$ for every m and $B_p \subset \bigcup_{m=1}^{\infty} B_{p_m}$. Applying our lemma to the norms p_m and q_m, we get $\mu((K \cap B_{p_m})^o) \geq 1 - 2\varepsilon$ for every m. Hence

$$\mu((K \cap B_p)^o) \geq \mu(\bigcap_{m=1}^{\infty} (K \cap B_{p_m})^o) = \lim_{m\to\infty} \mu((K \cap B_{p_m})^o) \geq 1 - 2\varepsilon. \blacksquare$$

(12.3) PROPOSITION. Let $(\Omega,\subset)$ be a directed set. For each $K \in \Omega$, let X_K be a Hausdorff topological space and let μ_K be a Radon measure on X_K with $\mu_K(X_K) = 1$. Suppose that, for each pair $(K,L) \in \Omega^2$ with $K \subset L$, a continuous mapping π_{KL} of X_L into X_K is given, such that

(1) $\pi_{KK} = id_{X_K}$ for each $K \in \Omega$,

(2) $\pi_{KL}\pi_{LM} = \pi_{KM}$

for each triple $(K,L,M) \in \Omega^3$ with $K \subset L \subset M$, and

(3) $\mu_K(A) = \mu_L(\pi_{KL}^{-1}(A))$

for each pair $(K,L) \in \Omega^2$ such that $K \subset L$ and for each $A \in \mathcal{B}(X_K)$. Suppose further that X is a Hausdorff topological space, and that, for each $K \in \Omega$, a continuous mapping π_K of X onto X_K is given, such that

(4) if $(K,L) \in \Omega^2$ and $K \subset L$, then $\pi_K = \pi_{KL}\pi_L$.

Finally, suppose that, for any two distinct points $x,y \in X$, there exists some $K \in \Omega$ such that $\pi_K(x) \neq \pi_K(y)$. Then the following two statements are equivalent:

(i) there is a unique Radon measure μ on X with $\mu(X) = 1$ such that $\mu_K(A) = \mu(\pi_K^{-1}(A))$ for each $K \in \Omega$ and each $A \in \mathcal{B}(X)$;

(ii) for each $\varepsilon > 0$, there is a compact subset Q of X such that $\mu_K(X_K \setminus \pi_K(Q)) \leq \varepsilon$ for each $K \in \Omega$.

This is Theorem 3.2 of [54].

(12.4) LEMMA. Let H be a subgroup of a nuclear vector group and ϕ a continuous p.d. function on H. Then there exists a finite regular Borel measure μ on $H_p^{\wedge}$ with $\hat{\mu} = \phi$.

Proof. Suppose first that $\phi(0) = 1$. Let K be a finitely generated subgroup of H. Then there exists on $K_p^\frown$ a unique Radon measure μ_K with $\hat{\mu}_K = \phi_{|K}$. Indeed, the closure $\overline{K}$ of K in span K, being a closed subgroup of a finite dimensional vector group, is locally compact. Hence, by the Weil-Raikov theorem, there is a Radon measure ν on $(\overline{K})_c^\frown$ with $\hat{\nu} = \phi_{|\overline{K}}$. We may identify $(\overline{K})_c^\frown$ with $K_{pc}^\frown$ because, evidently, each compact subset of $\overline{K}$ is contained in the closure of some precompact subset of K. Therefore we may treat ν as a measure on $K_{pc}^\frown$ and write $\hat{\nu} = \phi_{|K}$. The restriction μ_K of ν to Borel subsets of $K_p^\frown$ is then a Radon measure on $K_p^\frown$ with $\hat{\mu}_K = \phi_{|K}$. The uniqueness of μ_K is a consequence of (11.3). From (11.1) we get $\mu_K(K^\frown) = \phi(0) = 1$.

Let Ω be the directed family of all finitely generated subgroups of H. For each $K \in \Omega$, put $X_K = K_p^\frown$. For each pair $(K,L) \in \Omega^2$ such that $K \subset L$, let $\pi_{KL} : L_p^\frown \to K_p^\frown$ be the natural homomorphism. Put $X = H_p^\frown$ and, for each $K \in \Omega$, let $\pi_K : H_p^\frown \to K_p^\frown$ be the natural homomorphism.

In the situation we have just described, the assumptions of (12.3) are satisfied. The continuity of the homomorphisms π_K and π_{KL} is obvious and their surjectivity follows from (8.3). Conditions (1), (2) and (4) of (12.3) are satisfied trivially, as well as the fact that, for any two distinct $\chi_1, \chi_2 \in H^\frown$, there is some $K \in \Omega$ with $\pi_K(\chi_1) = \pi_K(\chi_2)$. Condition (3) is a consequence of (11.2). We shall prove that (ii) is satisfied.

Fix an $\varepsilon > 0$. Let F denote the nuclear vector group containing H. Since ϕ is continuous and $\phi(0) = 1$, there is some $U \in N_o(F)$ such that

$$\text{(1)} \qquad \operatorname{Re} \phi(h) \geq 1 - \frac{\varepsilon}{2} \qquad \text{for} \qquad h \in H \cap U.$$

Next, there are a linear subspace N of F and three pre-Hilbert seminorms p,q,r on N, such that $B_r \subset U$,

$$\text{(2)} \qquad 5 \sum_{k=1}^{\infty} k d_k(B_p, B_q) \leq \frac{1}{4},$$

$$\text{(3)} \qquad \sum_{k=1}^{\infty} d_k^{1/2}(B_q, B_r) < \frac{1}{12}$$

and $B_p \in N_o(F)$ (see (9.3), (2.14) and (2.15)). Then $W := H \cap B_p \in N_o(H)$, and W_H^o is a compact subset of $H_p^\frown$ (see (1.5)).

Take any $K \in \Omega$; we are to show that

(4) $$\mu_K(K^\frown \setminus \pi_K(W_H^o)) \leq \varepsilon.$$

Denote $L = K \cap N$ and $M = \operatorname{span} L$. Let s and t be the restrictions to M of q and r, respectively. Then s,t are pre-Hilbert seminorms on M and, by (2.13), we have

$$d_k(B_s,B_t) \leq d_k(B_q,B_r) \qquad (k = 1,2,\dots).$$

Hence, by (3),

$$\sum_{k=1}^{\infty} d_k^{1/2}(B_s,B_t) < \frac{1}{12}.$$

From (1) and (12.2) it now follows that

$$\mu_L((L \cap B_s)_L^o) \geq 1 - \varepsilon,$$

which may be written as

(5) $$\mu_K(\pi_{LK}^{-1}((L \cap B_q)_L^o)) \geq 1 - \varepsilon.$$

Since $K \cap B_q = L \cap B_q$, we have

(6) $$\pi_{LK}^{-1}((L \cap B_q)_L^o) \subset (K \cap B_q)_K^o.$$

From (2), (8.1) and (2.3) it follows that

(7) $$(K \cap B_q)_K^o \subset \pi_K(W_H^o).$$

Finally, from (5) - (7) we obtain (4) because $\mu_K(K^\frown) = \phi(0) = 1$.

We have proved that condition (ii) of (12.3) is satisfied. So, according to (12.3), there is a Radon measure μ on $H_p^\frown$ with $\mu(H_p^\frown) = 1$ such that $\mu_K(A) = \mu(\pi_K^{-1}(A))$ for each $K \in \Omega$ and each $A \in \mathcal{B}(K_p^\frown)$. For each $K \in \Omega$, the measure μ_K is the π_K-image of μ. Hence, by (11.2), we have $\hat{\mu}_{|K} = \hat{\mu}_K = \phi_{|K}$. Thus $\hat{\mu} = \phi$. It remains to show that μ is a regular measure.

As we have just seen, to each n there corresponds some $W_n \in \mathcal{N}_o(H)$ such that $\mu_K(K^\frown \setminus \pi_K(W_n^o)) \leq \frac{1}{n}$ for all $K \in \Omega$. The set $Q_n = W_n^o$ is equicontinuous and, by (1.5), compact in $H_p^\wedge$. The family of sets of the form $\pi_K^{-1}(K_p^\wedge \setminus \pi_K(Q_n))$ where $K \in \Omega$ forms an open covering of $H_p^\wedge \setminus Q_n$. Hence, for each compact $Y \subset H_p^\wedge \setminus Q_n$, we can find some $K \in \Omega$ such that $Y \subset \pi_K^{-1}(K^\wedge \setminus \pi_K(Q_n))$. Then

$$\mu(Y) \leq \mu(\pi_K^{-1}(K^\wedge \setminus \pi_K(Q_n))) = \mu_K(K^\wedge \setminus \pi_K(Q_n)) \leq \frac{1}{n}.$$

This implies that $\mu(H^\wedge \setminus Q_n) \leq \frac{1}{n}$ since μ is a Radon measure. Thus μ is regular.

It remains to consider the case $\phi(0) \neq 1$. If $\phi(0) \neq 0$, then it suffices to consider the p.d. function $\phi/\phi(0)$ instead of ϕ. On the other hand, if $\phi(0) = 0$, then, by (1.22) (b), we have $\phi \equiv 0$ and ϕ is the transform of the zero measure. ▪

Proof of (12.1). Due to (11.1), if μ is a regular finite Borel measure on $G_\tau^\wedge$, then $\hat{\mu}$ is a continuous p.d. function on G. From (11.3) it follows that the mapping $\mu \to \hat{\mu}$ is injective. We shall prove that it is surjective, as well.

So, take any continuous p.d. function ϕ on G. By (9.6) there exist a nuclear vector group F, a subgroup H of F and a closed subgroup K of H, such that $G \sim H/K$. We may simply assume that $G = H/K$. Let $\psi : H \to H/K$ be the natural projection. Then $\phi\psi$ is a continuous p.d. function on H and, due to (12.4), there exists a regular finite Borel measure ν on $H_p^\wedge$ with $\hat{\nu} = \phi\psi$. Let $\pi : H_p^\wedge \to K_p^\wedge$ be the natural homomorphism. By (11.2), we have $\hat{\nu}_\pi = \hat{\nu}_{|K} = \phi\psi_{|K} \equiv \phi(0)$. In view of (11.3), this means that ν_π is the Dirac measure. Then $\nu(K^o) = \nu(\pi^{-1}(0)) = \nu_\pi(\{0\}) = \phi(0)$, i.e. ν is concentrated on K^o. We may treat ν as a finite regular Borel measure on K_p^o.

The natural homomorphism $\sigma : K_p^o \to (H/K)_p^\wedge$ is continuous (in fact, it is a topological isomorphism; however, the mapping $\sigma : K_{pc}^o \to (H/K)_c^\wedge$ need not be continuous; consider e.g. the canonical mapping $L_{pc}^o \to (\mathcal{D}/L)_c^\wedge$ in (17.6)). Set $\lambda = \nu_\sigma$. Then λ is a finite regular Borel measure on $G_p^\wedge$. For each $h \in H$, one has

$$\hat{\lambda}(\psi(h)) = \int_{G_p^\wedge} \chi(\psi(h))d\lambda(\chi) = \int_{K_p^o} \kappa(h)d\nu(\kappa) = \int_{H_p^\wedge} \kappa(h)d\nu(\kappa)$$
$$= \hat{\nu}(h) = \phi(\psi(h)),$$

which proves that $\hat{\lambda} = \phi$.

Since ν is regular and $\nu(H_p^{\wedge}) = \phi(0)$, there is a sequence $U_n \in N_o(H)$ with $\nu(U_n^o) \to \phi(0)$. The sets $Q_n = (\psi(U_n))^o$ are equicontinuous and compact in $G_\tau^{\wedge}$. Moreover,

$$\lambda(Q_n) = \lambda(\psi(U_n)^o) = \lambda(\sigma(U_n^o \cap K^o)) = \nu(U_n^o \cap K^o) = \nu(U_n^o) \to \phi(0)$$
$$= \lambda(G_p^{\wedge}),$$

i.e. $\lambda(G_p^{\wedge} \setminus Q_n) \to 0$. Let $\overline{\lambda}$ be the completion of λ. Take any closed subset Y of $G_\tau^{\wedge}$. The sets $Y \cap Q_n$, $n = 1,2,\ldots,$ are compact in $G_\tau^{\wedge}$, thus in $G_p^{\wedge}$. Hence $\bigcup_{n=1}^{\infty} (Y \cap Q_n) \in B(G_p^{\wedge})$. On the other hand,

$$\lambda(Y \setminus \bigcup_{n=1}^{\infty} (Y \cap Q_n)) \leq \lambda(\bigcap_{n=1}^{\infty} (G_p^{\wedge} \setminus Q_n)) = 0.$$

This implies that Y is $\overline{\lambda}$-measurable. Hence all Borel subsets of $G_\tau^{\wedge}$ are $\overline{\lambda}$-measurable.

Let μ be the restriction of $\overline{\lambda}$ to $B(G_\tau^{\wedge})$. It is clear that $\hat{\mu} = \hat{\lambda}$. Hence $\hat{\mu} = \phi$. It remains to prove that μ is regular. So, take any $A \in B(G_\tau^{\wedge})$ and $\varepsilon > 0$. Since A is $\overline{\lambda}$-measurable, there are some $B,N \in B(G_p^{\wedge})$ with $\lambda(N) = 0$ and $B \setminus N \subset A \subset B \cup N$. Next, since λ is regular, there exists a compact equicontinuous subset Q of $B \setminus N$ with $\lambda(B \setminus N \setminus Q) < \varepsilon$. Take n so large that $\lambda(G_p^{\wedge} \setminus Q_n) < \varepsilon$. The set Q is closed in $G_p^{\wedge}$, thus in $G_\tau^{\wedge}$. Therefore $Q \cap Q_n$ is compact in $G_\tau^{\wedge}$. Finally, we have

$$\lambda(A \setminus (Q \cap Q_n)) = \lambda(B \setminus (Q \cap Q_n)) \leq \lambda(B \setminus Q) + \lambda(B \setminus Q_n)$$
$$< \varepsilon + \varepsilon. \quad \blacksquare$$

(12.5) THEOREM. Let G be a nuclear group and τ an admissible topology on $G^{\wedge}$. Let $\{\mu_i\}_{i\in I}$ be a family of regular Borel measures on $G_\tau^{\wedge}$, with $\mu_i(G^{\wedge}) = 1$ for every i. Then the family $\{\hat{\mu}_i\}_{i\in I}$ of p.d. functions on G is equicontinuous at zero if and only if to each $\varepsilon > 0$ there corresponds a compact equicontinuous subset Q of $G_\tau^{\wedge}$ such that $\mu_i(G^{\wedge} \setminus Q) \leq \varepsilon$ for all $i \in I$.

Proof. The sufficiency of the condition follows immediately from the proof of (11.1). We shall prove the necessity. It is not hard to see that, without loss of generality, we may assume that τ is the topology of pointwise convergence. Next, due to (9.6), we may assume that $G = H/K$

where H is a subgroup of some nuclear vector group F, and K is a closed subgroup of H. Suppose first that $K = \{0\}$.

Choose any $\varepsilon > 0$. We can find some $U \in \mathcal{N}_0(F)$ such that

(1) $$\operatorname{Re} \hat{\mu}_i(u) \geq 1 - \frac{\varepsilon}{2} \quad \text{for all} \quad u \in U \quad \text{and} \quad i \in I.$$

According to (9.3) and (2.14), we can find some linear subspace N of F and two pre-Hilbert seminorms p,q on N, such that $B_p \in \mathcal{N}_0(F)$,

(2) $$\sum_{k=1}^{\infty} d_k^{1/2}(B_p,B_q) < \frac{1}{12}$$

and $B_q \subset U$. Take any $i \in I$ and any compact $X \subset H_p^{\wedge} \setminus (H \cap B_p)_H^{\circ}$. Since μ_i is a regular measure, it is enough to show that $\mu_i(X) \leq \varepsilon$.

Let Ω be the family of all finitely generated subgroups of H. For each $L \in \Omega$, let $\pi_L : H_p^{\wedge} \to L_p^{\wedge}$ be the natural homomorphism. It is clear that

$$(H \cap B_p)_H^{\circ} = \bigcap_{L \in \Omega} \pi_L^{-1}((L \cap B_p)_L^{\circ}),$$

whence

$$H^{\wedge} \setminus (H \cap B_p)_H^{\circ} = \bigcup_{L \in \Omega} \pi_L^{-1}[L^{\wedge} \setminus (L \cap B_p)_L^{\circ}].$$

Since X is compact, it follows that there is some $L \in \Omega$ with $X \subset \pi_L^{-1}[L^{\wedge} \setminus (L \cap B_p)_L^{\circ}]$. Let μ_L be the π_L-image of μ_i. From (11.2) we get $\hat{\mu}_{i|L} = \hat{\mu}_L$. Hence, by (1), (2), (12.2) and (2.13), we obtain

$$\mu_i(X) \leq \mu_i[\pi_L^{-1}(L^{\wedge} \setminus B_p)_L^{\circ}] = \mu_L(L^{\wedge} \setminus (L \cap B_p)_L^{\circ}) \leq \varepsilon.$$

It remains to consider the case when $K \neq \{0\}$. Let $\psi : H \to H/K$ be the canonical projection and $\phi : (H/K)_p^{\wedge} \to H_p^{\wedge}$ the dual homomorphism. For every i, let ν_i be the ϕ-image of μ_i; then $\hat{\nu}_i = \hat{\mu}_i \psi$ according to (11.2). Since the family $\{\hat{\nu}_i\} = \{\hat{\mu}_i \psi\}$ of p.d. functions on H is equicontinuous at zero, the above implies that to each $\varepsilon > 0$ there corresponds a compact equicontinuous subset Q of $H_p^{\wedge}$ with $\nu_i(H^{\wedge} \setminus Q) \leq \varepsilon$ for all $i \in I$. Then

$$\mu_i((H/K)^{\wedge} \setminus \phi^{-1}(Q)) = \mu_i(\phi^{-1}(H^{\wedge} \setminus Q)) = \nu_i(H^{\wedge} \setminus Q) \leq \varepsilon.$$

Now, it remains to observe that the closure of $\phi^{-1}(Q)$ is a compact equicontinuous subset of $(H/K)_p^{\wedge}$ (cf. (1.5)). ■

(12.6) NOTE. The material of this section is now.

13. The SNAG theorem

(13.1) LEMMA. Let G be an abelian topological group and τ a topology on $G^\frown$ such that the mappings $G^\frown_\tau \ni \kappa \to \kappa(g)$, $g \in G$, are continuous. Let $\alpha : G \to (G^\frown_\tau)^\frown$ be the natural homomorphism and μ a Radon measure on $G^\frown_\tau$. Then span $\alpha(G)$ is a dense subset of $L^2_C(G^\frown_\tau, \mu)$.

Proof. It is enough to show that the characteristic function of compact subsets of $G^\frown_\tau$ can be approximated in $L^2_C(G^\frown_\tau, \mu)$ by elements of span $\alpha(G)$. So, take any compact subset X of $G^\frown_\tau$ and any $\varepsilon > 0$. Since μ is a Radon measure, there exists a compact subset Y of $G^\frown_\tau \setminus X$ with

(1) $\mu(G^\frown \setminus X \setminus Y) < \varepsilon.$

Denote $U = G^\frown \setminus Y$. Since Y is a compact subset of $G^\frown_\tau$, it is a compact subset of $G^\frown_p$, which means that U is an open subset of $G^\frown_p$. So, for each $\xi \in X$, there are a finite subset G_ξ of G and some $\delta_\xi > 0$, such that

(2) $U_\xi := \{\kappa \in G^\frown : |\kappa(g) - \xi(g)| < \delta_\xi \quad \text{for all} \quad g \in G_\xi\} \subset U.$

The sets U_ξ, $\xi \in X$, are open in $G^\frown_p$ and cover the compact set X. So, there is a finite subset Q of X such that

(3) $X \subset \bigcup_{\xi \in Q} U_\xi.$

Arrange all elements of the finite set $\bigcup_{\xi \in Q} G_\xi$ in a sequence $(g_k)^n_{k=1}$. Let $\phi : G^\frown_p \to S^n$ be the continuous homomorphism given by the formula $\phi(\chi) = (\kappa(g_k))^n_{k=1}$ for $\kappa \in G^\frown$. Denote $V = \phi^{-1}(\phi(X))$. We shall prove that

(4) $X \subset V \subset U.$

The first inclusion is trivial. To prove the second one, choose any $\kappa \in V$. We have $\phi(\kappa) = \phi(\chi)$ for some $\chi \in X$. Next, by (3), we have $\chi \in U_\xi$ for some $\xi \in Q$. This implies that $\kappa \in U_\xi$. Hence, by (2), we get $\kappa \in U$, which proves (4).

The restriction ν of μ to Borel subsets of $G_p^\wedge$ is a Radon measure on $G_p^\wedge$. So, ν_ϕ is a Radon measure on S^n. Let $\| \ \|_{G^\wedge}$ and $\| \ \|_{S^n}$ denote the norms in the spaces $L^2_C(G^\wedge,\mu)$ and $L^2_C(S^n,\nu_\phi)$, respectively. Let χ_A denote the characteristic function of a set A. Since trigonometric polynomials are dense in $L^2_C(S^n,\nu_\phi)$ (see [38], (31.4)), we can find some complex numbers $\lambda_1,\ldots,\lambda_p$ and some continuous homomorphisms (characters) $\eta_k : S^n \to S$, $k = 1,\ldots,p$, such that

$$\|\chi_{\phi(X)} - \sum_{k=1}^{p} \lambda_k \eta_k\|_{S^n} < \varepsilon.$$

Then

$$\|\chi_V - \sum_{k=1}^{p} \lambda_k\eta_k\phi\|_{G^\wedge} = [\int_{G^\wedge} |\chi_V - \sum_{k=1}^{p} \lambda_k\eta_k\phi|^2 d\nu]^{1/2}$$

$$= [\int_{S^n} |\chi_{\phi(X)} - \sum_{k=1}^{p} \lambda_k\eta_k|^2 d\nu_\phi]^{1/2} < \varepsilon.$$

Hence, applying (1) and (4), we obtain

$$\|\chi_X - \sum_{k=1}^{p} \lambda_k\eta_k\phi\|_{G^\wedge} \leq \|\chi_X - \chi_V\|_{G^\wedge} + \|\chi_V - \sum_{k=1}^{p} \lambda_k\eta_k\phi\|_{G^\wedge}$$

$$\leq [\mu(U \setminus X)]^{1/2} + \varepsilon < \varepsilon^{1/2} + \varepsilon.$$

Thus it remains to show that $\eta_k\phi \in \alpha(G)$ for $k = 1,\ldots,p$.

Fix any $k = 1,\ldots,p$. We can find some $s_1,\ldots,s_n \in Z$ such that

$$\eta_k(z_1,\ldots,z_n) = z_1^{s_1} \ldots z_n^{s_n}$$

for $(z_1,\ldots,z_n) \in S^n$ (cf. (1.7)). Then, for each $\kappa \in G^\wedge$, we have

$$(\eta_k\phi)(\kappa) = \eta_k(\kappa(g_1),\ldots,\kappa(g_n)) = \prod_{k=1}^{n} \kappa(g_k)^{s_k} = \kappa(\sum_{k=1}^{n} s_k g_k),$$

which means that $\eta_k\phi = \alpha(\sum_{k=1}^{n} s_k g_k) \in \alpha(G)$. ∎

(13.2) **THEOREM.** Let Φ be a continuous cyclic unitary representation of a nuclear group G and let τ be an admissible topology on

$G^\wedge$. Then there exists on $G^\wedge_\tau$ a regular finite Borel measure μ such that Φ is unitarily equivalent to the representation Ψ in the space $L^2_C(G^\wedge_\tau,\mu)$, given by the formula

$$(\Psi_g f)(\chi) = f(\chi)\ \exp\,[2\pi i\chi(g)]$$

for $g \in G$, $\chi \in G^\wedge$ and $L^2_C(G^\wedge_\tau,\mu)$.

Proof. Let u be a cyclic vector of Φ. Then the formula $\phi(g) = (\Phi_g u,u)$, $g \in G$, defines a continuous p.d. function ϕ on G (see (1.23)). By (12.1), there exists a regular finite Borel measure μ on $G^\wedge_\tau$ with $\hat{\mu} = \phi$. Let f be the function identically equal to 1 on $G^\wedge$. From (13.1) it follows that f is a cyclic vector of Ψ. For each $g \in G$, we have

$$(\Psi_g f,f) = \int_{G^\wedge} \exp\,[2\pi i\chi(g)]d\mu(\chi) = \hat{\mu}(g) = \phi(g) = (\Phi_g u,u).$$

This implies that Φ and Ψ are unitarily equivalent (see (1.25)). ∎

Since a continuous unitary representation can be written as the Hilbert sum of cyclic representation (see (1.21)), we could easily formulate an analogue of (13.2) for non-cyclic representations (cf.[38], (21.14)). Instead, we shall give the SNAG theorem for nuclear group in its usual form (cf. [18], p. 160).

(13.3) THEOREM. Let Φ be a continuous unitary representation of a nuclear group G and let τ be an admissible topology on $G^\wedge$. Then there exists on Borel subsets of $G^\wedge_\tau$ a unique spectral measure P such that

$$(1) \qquad \Phi_g = \int_{G^\wedge} \exp\,[2\pi i\chi(g)]dP(\chi) \qquad (g \in G).$$

For the definition of the spectral measure, see [18], app. B. 3.

Proof. In view of (1.21), we may assume that Φ is the Hilbert sum of some continuous cyclic unitary representations Φ_i, $i \in I$. Next, according to (13.2), we may assume that, for each $i \in I$, there exists a regular finite Borel measure μ_i on $G^\wedge_\tau$ such that Φ_i is the representation in the space $L^2_C(G^\wedge_\tau,\mu_i)$, given by the formula

$$(\Phi_i(g)f)(\chi) = f(\chi)\ \exp\,[2\pi i\chi(g)]$$

for $g \in G$, $\chi \in G^{\wedge}$ and $f \in L^2_C(G^{\wedge}_\tau, \mu_i)$. Let $H = \bigoplus_{i \in I} L^2_C(G^{\wedge}_\tau, \mu_i)$. For each $A \in \mathcal{B}(G^{\wedge}_\tau)$, we define a projection $P(A) : H \to H$ by

$$P(A)(f_i)_{i\in I} = (\chi_A f_i)_{i\in I}$$

where $f_i \in L^2_C(G^{\wedge}_\tau, \mu_i)$ for every i and χ_A denotes the characteristic function of A. It is not difficult to verify that P is a spectral measure on H, satisfying (1).

To prove the uniqueness of P, suppose that we are given two spectral measures P_1 and P_2 defined on Borel subsets of $G^{\wedge}_\tau$ such that

$$\Phi_g = \int_{G^{\wedge}} \exp\,[2\pi i\chi(g)]dP_i(\chi) \qquad (g \in G,\ \ i = 1,2,).$$

For each $h \in H$, the mapping

$$\mathcal{B}\,(G^{\wedge}_\tau) \ni A \to P^h_i(A) : = (P_i(A)\,h,h) \qquad (i = 1,2),$$

is a regular finite Borel measure on $G^{\wedge}_\tau$, and

$$(\Phi_g h,h) = \int_{G^{\wedge}} \exp\,[2\pi i\chi(g)]dP^h_i \qquad (g \in G,\ \ i = 1,2).$$

Hence, in view of the uniqueness of the measure in (12.1), we have $P^h_1 = P^h_2$. Since h was arbitrary, it follows that $P_1 = P_2$. ∎

We shall now deal with the problem of the extending of continuous p.d. functions and unitary representations.

(13.4) LEMMA. Let H be a subgroup of a nuclear vector group F. Then each continuous unitary representation of H can be extended to a continuous unitary representation of F.

Proof. Let Φ be a continuous unitary representation of H. According to (1.21), we may write Φ as the Hilbert sum of some continuous cyclic unitary representations Φ_ν. Suppose that, for every ν, we have found a continuous unitary representation Ψ_ν of F with $\Psi_{\nu|H} = \Psi_\nu$. Then $\Psi = \bigoplus \Psi_\nu$ is a continuous unitary representation of F with $\Psi_{|H} = \Phi$. Therefore, Φ may be assumed to be a cyclic representation. Then, in virtue of of (13.2), we may assume that there exists on $H^{\wedge}_p$ a regular finite Borel measure μ such that Φ is the representation in the space $L^2_C(H^\wedge,\mu)$, given by the formula

$$(\Phi_h f)(\chi) = f(\chi) \exp [2\pi i\chi(h)]$$

for $h \in H$, $\chi \in H^\wedge$ and $f \in L^2_C(H^\wedge,\mu)$.

For each $n = 1,2,\ldots,$ there is a compact equicontinuous subset Q_n of $H^\wedge_p$ with $\mu(H^\wedge \setminus Q_n) < 1/n$. We may assume that $Q_n \subset Q_{n+1}$. Let L^2_n be the subspace of $L^2_C(H^\wedge,\mu)$ consisting of functions with support in $Q_n \setminus Q_{n-1}$. Then $L^2_C(H^\wedge,\mu)$ is the Hilbert sum of the invariant subspaces L^2_n and Φ is the Hilbert sum of the corresponding representations Φ_n. As before, we have to show only that every Φ_n can be extended to a continuous unitary representation Ψ_n of F. Therefore, we may simply assume that there is a compact equicontinuous subset Q of $H^\wedge_p$ with $\mu(H^\wedge \setminus Q) = 0$. Then there is some $U \in N_o(F)$ with $Q \subset (U \cap H)^o_H$. As in the proof of (12.4), we can find a linear subspace N of F and three pre-Hilbert seminorms p,q,r of N, such that $B_p \in N_o(F)$,

(1) $$\lim_{k\to\infty} d_k(B_p,B_q) = 0,$$

(2) $$5 \sum_{k=1}^{\infty} k d_k(B_q,B_r) \le 1,$$

and $B_r \subset U$. We may assume that $N = F$. Indeed, if Ψ' is a continuous unitary representation of N with $\Psi' = \Phi$ on $H \cap N$, then the formula

$$\Psi''(h + u) = \Phi(h)\Psi'(u) \qquad (h \in H,\ u \in N)$$

defined a unitary representation Ψ'' of $H + N$. Since the group of unitary operators is divisible, we can extend Ψ'' to a unitary representation Ψ of F (cf. (1.6)). We have $\Psi_{|H} = \Psi''_{|H} = \Phi$. Moreover, $\Psi_{|N} = \Psi''_{|N} = \Psi'$. Since Ψ' is continuous and N is an open subgroup of F, it follows that Ψ is continuous.

Consider the canonical diagram

$$\begin{array}{ccccc} F & \xrightarrow{\ \ id\ \ } & F & \xrightarrow{\ \ id\ \ } & F \\ \Big\downarrow \psi_p & & \Big\downarrow \psi_d & & \Big\downarrow \\ F_p & \xrightarrow{\ \ \Lambda_{pq}\ \ } & F_q & \xrightarrow{\ \ \Lambda_{qr}\ \ } & F_r \end{array}$$

Let B be the closed unit ball of $(F_q)^*$. It follows from (1) that $\Lambda_{\rho q}$ is a compact operator, which implies that F_q is separable. Thus B is a Polish space.

Consider the relation

$$\Xi = \{(\chi,f) \in Q \times B : \rho f \psi_q(h) = \chi(h) \quad \text{for all} \quad h \in H\}.$$

We shall prove that Ξ is a multifunction from Q to B (for the terminology concerning multifunctions, cf. [40]). So, take any $\chi \in Q$. The formula

$$\chi'(h + u) = \chi(h) \qquad\qquad (h \in H, \quad u \in r^{-1}(0))$$

defines a continuous character χ' of the group $H' = H + r^{-1}(0)$. We have $|\chi'(H' \cap B_r)| \leq 1/4$. By (2) and (8.1), there exists some $f' \in F^\#$ with $\rho f'_{|H'} = \chi'$ and $\sup\{|f'(u)| : u \in B_q\} \leq 1$. Since $\chi' \equiv 0$ on $r^{-1}(0)$, it follows that $f' \equiv 0$ on $q^{-1}(0)$. So, there is some $f \in B$ with $f' = f\psi_q$. Then $(\chi,f) \in \Xi$.

It is obvious that the graph of Ξ is closed. Hence, by Aumann's theorem on measurable selectors (see [40], Theorem 5.2), there exists a Borel mapping $\alpha : Q \to B$ such that $\rho\alpha(\chi)\psi_{q|H} = \chi$ for μ-almost all $\chi \in Q$. Consider the continuous mapping $\beta : B \to F_p^\wedge$ given by the formula $\beta(f) = \rho f \psi_q$ for $f \in B$. The set $\beta(B)$ is equicontinuous because, evidently, $\beta(B) \subset (\frac{1}{4} B_q)^o_F$. Next, let $\sigma : H_p^\wedge \to F_p^\wedge$ be the Borel mapping given by the formula

$$\sigma(\chi) = \begin{cases} \beta\alpha(\chi) & \text{if} \quad \chi \in Q \\ 0 & \text{if} \quad \chi \notin Q. \end{cases}$$

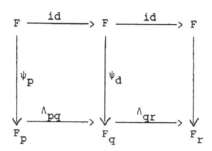

The image of σ is equicontinuous because $\sigma(H^{\wedge}) = \sigma(Q) \subset \beta(B)$. Therefore the Fourier transform of the Borel measure μ_σ is continuous (cf. the proof of (11.1)). Let Ψ' be the unitary representation of F in the space $L^2_C(F^{\wedge},\mu_\sigma)$, given by the formula

$$(\Psi'_u f)(\kappa) = f(\kappa)\ \exp\,[2\pi i\kappa(u)]$$

for $u \in F$, $f \in L^2_C(F^{\wedge},\mu_\sigma)$ and $\kappa \in F^{\wedge}$. Due to (13.1), the function $f_o \equiv 1$ is a cyclic vector of Ψ'. We have

$$(\Psi'_u f_o, f_o) = \int_{F^{\wedge}} \exp\,[2\pi i\kappa(u)]d\mu_\sigma(\kappa) = \hat{\mu}_\sigma(u)$$

for all $u \in F$, which means that Ψ' is continuous (see (1.23)).

For almost all $\chi \in Q$, we have

$$\sigma(\chi)_{|H} = (\beta\alpha)(\chi)_{|H} = \rho\alpha(\chi)\psi_{q|H} = \chi,$$

which implies that

(3) $\qquad \sigma(\chi)_{|H} = \chi \quad$ for almost all $\chi \in H^{\wedge}$.

Hence, for each $f \in L^2_C(H^{\wedge},\mu)$, we have

$$\int_{F^{\wedge}} |f(\kappa_{|H})|^2 d\mu_\sigma(\kappa) = \int_{H^{\wedge}} |f(\sigma(\chi)_{|H})|^2 d\mu(\chi) = \int_{H^{\wedge}} |f(\chi)|^2 d\mu(\chi).$$

This means that the formula

$$(\Gamma f)(\kappa) = f(\kappa_{|H}) \qquad (\kappa \in F^{\wedge})$$

defines an isometric operator $\Gamma : L^2_C(H^{\wedge},\mu) \to L^2_C(F^{\wedge},\mu_\sigma)$. On the other hand, for each $f \in L^2_C(F^{\wedge},\mu_\sigma)$, we have

$$\int_{H^{\wedge}} |f(\sigma(\chi))|^2 d\mu(\chi) = \int_{F^{\wedge}} |f(\kappa)|^2 d\mu_\sigma(\kappa),$$

which means that the formula

$$(\Delta f)(\chi) = f(\sigma(\chi)) \qquad (\chi \in H^{\wedge})$$

defines an isometric operator $\Delta : L^2_C(F^{\wedge},\mu_\sigma) \to L^2_C(H^{\wedge},\mu)$.

Take any $f \in L^2_C(H^{\wedge},\mu)$. According to (3), we have

$$\Delta\Gamma f(\chi) = (\Gamma f)(\sigma(\chi)) = f(\sigma(\chi)_{|H}) = f(\chi)$$

for almost all $\chi \in H^{\wedge}$. Thus $\Delta\Gamma = \mathrm{id}$, which implies that Γ is an isometry of $L^2_C(H^{\wedge},\mu)$ onto $L^2_C(F^{\wedge},\mu_\sigma)$. Then $\Psi := \Gamma^{-1}\Psi'\Gamma$ is a continuous unitary representation of F in the space $L^2_C(H^{\wedge},\mu)$.

Take any $h \in H$ and $f \in L^2_C(H^{\wedge},\mu)$. In view of (3), for almost all $\chi \in H^{\wedge}$, we have

$$(\Psi_h f)(\chi) = (\Delta\Psi'_h\Gamma f)(\chi) = (\Psi'_h\Gamma f)(\sigma(\chi))$$

$$= (\Gamma f)(\sigma(\chi)) \exp[2\pi i\sigma(\chi)(h)] = f(\sigma(\chi)_{|H}) \exp[2\pi i\chi(h)]$$

$$= f(\chi) \exp[2\pi i\chi(h)] = (\Phi_h f)(\chi).$$

This proves that $\Psi_{|H} = \Phi$. ∎

(13.5) THEOREM. Let A be a subgroup of a nuclear group G. Then each continuous unitary representation of A can be extended to a continuous unitary representation of G.

Proof. Let Φ be a continuous unitary representation of A. In virtue of (9.6), we may assume that there exist a nuclear vector group F, a subgroup H of F and a closed subgroup K of H, such that $G = H/K$. Let $\psi : H \to H/K$ be the canonical projection. Then $\Phi\psi$ is a continuous unitary representation of $\psi^{-1}(A)$. By (13.4), we can extend it to a continuous unitary representation Ψ' of H. Since Ψ' is trivial on K, it induces a continuous unitary representation Ψ of H/K with $\Psi\psi = \Psi'_{|H}$. It is obvious that $\Psi_{|A} = \Phi$. ∎

(13.6) THEOREM. Let A be a subgroup of a nuclear group G. Then each continuous p.d. function on A can be extended to a continuous p.d. function on G.

Proof. Let ϕ be a continuous p.d. function on A. There exists a continuous cyclic unitary representation Φ of A with a cyclic vector u, such that $(\Phi_g u,u) = \phi(g)$ for all $g \in A$ (see (1.24)). By (13.5), we can extend Φ to a continuous unitary representation Ψ of G. Then the function $G \ni g \to (\Psi_g u,u)$ is a continuous p.d. extension of ϕ (see (1.23)). ∎

(13.7) NOTE. The material of this section is new.

Chapter 5

PONTRYAGIN DUALITY

Let K be an additive subgroup of a real locally convex space E. Let E^* be the dual space and K^* the subgroup of E^* consisting of functionals which assume integer values at points belonging to K. We saw in section 8 that closed subgroups of nuclear spaces are weakly closed. Thus, if K is a closed subgroup of a nuclear Fréchet space, we may write $K^{**} = K$. The equality $K^{**} = K$ for closed subgroups of R^n is the heart of the Pontryagin duality for LCA groups. These facts were the starting point for the investigation of duality properties of additive subgroups and quotients of nuclear spaces.

Section 14 of this chapter contains the basic facts on duality for subgroups, quotient groups, products and direct sums of abelian topological groups. In section 15 we give an account of known results on the Pontryagin duality for locally convex spaces. In section 16 we prove that character groups of metrizable nuclear (vector) groups are nuclear. Section 17 contains the proof of the main result of this chapter: if G is a countable product of LCA groups and metrizable, complete nuclear groups, then the duality between G and $G^\wedge$ induces dualities between appropriate closed subgroups and quotients of G and $G^\wedge$. We also give here some examples which show that the assumptions of countability and metrizability are essential. Finally, section 18 is devoted to Vilenkin's theory of groups with boundedness.

Let H be a subgroup of an abelian topological group G. To simplify the notation, in this chapter we shall assume that the groups $G^\wedge$ and H^o are endowed with the compact-open topology. In some places, however, where the difference between various topologies on $G^\wedge$ is essential, we shall apply the notation $G^\wedge_c, G^\wedge_{pc}$ etc.

14. Preliminaries

Let G be an abelian topological group. The evaluation map from G into $G^{\wedge\wedge}$ will be denoted by α_G. Thus $\langle\alpha_G(g)\rangle = \langle\chi, g\rangle$ for $g \in G$, $\chi \in G^\wedge$. Obviously, α_G is injective if and only if G admits sufficiently many continuous characters. We say that G is a <u>reflexive group</u> if α_G is a topological isomorphism of G onto $G^{\wedge\wedge}$.

(14.1) LEMMA. Let A be an arbitrary subset of an abelian topological group G and let Q be the quasi-convex hull of A. Then $\alpha_Q(Q) = \alpha_G(G) \cap A^{oo}$.

Proof. The inclusion $\alpha_G(Q) \subset \alpha_G(G) \cap A^{oo}$ is trivial. To prove the opposite one, choose an arbitrary $\zeta \in \alpha_G(G) \cap A^{oo}$. We have $\zeta = \alpha_G(g)$ for a certain $g \in G$ and it remains to show that $g \in Q$. Suppose the contrary; then $|\chi(g)| > \frac{1}{4}$ for a certain $\chi \in A^o$. Hence $|\zeta(\chi)| = |\chi(g)| > \frac{1}{4}$, which is impossible because $\zeta \in A^{oo}$. ∎

(14.2) LEMMA. Let G be an abelian topological group with $\alpha_G(G) = G^{\wedge\wedge}$. For each dually closed subgroup H of G, one has $\alpha_G(H) = H^{oo}$. Each dually closed subgroup of $G^\wedge$ is the annihilitor of some closed subgroup of G.

Proof. The first assertion is a direct consequence of (14.1) because dually closed subgroups are quasi-convex sets. To prove the second one, take a dually closed subgroup A of $G^\wedge$. The set

$$B = \{g \in G : \chi(g) = 0 \quad \text{for all} \quad \chi \in A\}$$

is a closed subgroup of G. It is obvious that $A \subset B^o$. So, it remains to show that $B^o \subset A$.

Suppose the contrary and let $\kappa \in B^o \setminus A$. Since A is dually closed in $G^\wedge$, we have $|\zeta(\kappa)| > \frac{1}{4}$ for a certain $\zeta \in A^o$. Next, we have $\zeta = \alpha_G(g)$ for a certain $g \in G$ because $\alpha_G(G) = G^{\wedge\wedge}$. Since $\zeta \in A^o$, it follows that $\chi(g) = \zeta(\chi) = 0$ for all $\chi \in A$, whence $g \in B$. But $\kappa \in B^o$ and, therefore, $\kappa(g) = 0$, which is impossible because $\kappa(g) = \zeta(\kappa)$. ∎

(14.3) LEMMA. Let G be a locally quasi-convex group and let G' be the image of α_G endowed with the topology induced from $G^{\wedge\wedge}$. Then the mapping $\alpha_G : G \to G'$ is open.

Proof. Choose an arbitrary $U \in \mathcal{N}_o(G)$. We have to show that $\alpha_G(U) \in \mathcal{N}_o(G')$. We may assume U to be quasi-convex. Then, by (14.1), we have $\alpha_G(U) = G' \cap U^{oo}$ and it remains to observe that $U^{oo} \in \mathcal{N}_o(G^{\wedge\wedge})$ because U^o is a compact subset of $G^\wedge$ (see (1.5)). ∎

(14.4) LEMMA. If an abelian topological group G is a k-space, then α_G is continuous.

This follows immediately from the fact that, due to the Ascoli theorem, compact subsets of $G^{\wedge}$ are equicontinuous.

(14.5) REMARK. Let X,Y be topological spaces. A mapping $f : X \to Y$ is called k-continuous if all its restrictions to compact subsets of X are continuous. A topological group G is called a k-group if all k-continuous homomorphisms of G into topological groups are continuous. This notion was introduced by N. Noble [72]. Naturally, if G is a k-space, it is also a k-group, but there exist k-groups which are not k-spaces. Noble proved that if an abelian topological group G is a k--group, then α_G is continuous ([72], Theorem 2.3).

For our purposes, we get little benefit from the notion of a k--group. An abelian topological group G will be called a c-group if α_G is continuous. It turns out that the permanence properties of k--groups, established in [72], remain valid for c-groups. Consequently, if we are able to prove that a given topological group is a k-group, then we are also in a position to prove that it is a c-group, without referring to the notion of a k-group. Besides, there exist c-groups which are not k-groups, e.g. uncountable direct sums of real lines.

(14.6) LEMMA. Let H be a dually closed, dually embedded subgroup of an abelian topological group G. If $\alpha_G(G) = G^{\wedge\wedge}$, then $\alpha_H(H) = H^{\wedge\wedge}$.

Proof. Take an arbitrary $\xi \in H^{\wedge\wedge}$ and consider the canonical homomorphism $\phi : G^{\wedge} \to H^{\wedge}$. It is obvious that $\xi\phi \in H^{oo}$. By (14.2), we have $\alpha_G(H) = H^{oo}$. Consequently, there is some $h \in H$ with $\alpha_G(h) = \xi\phi$ and it remains to show that $\alpha_H(h) = \xi$. For each $\chi \in G^{\wedge}$, one has

$$\langle\alpha_H(h),h\chi\rangle = \langle\alpha_H(h),\phi(\chi)\rangle = \langle\alpha_H(h),\chi_{|H}\rangle = \langle\chi,h\rangle$$

$$= \langle\alpha_G(h),\chi\rangle = \langle\xi\phi,\chi\rangle.$$

This means that $\alpha_H(h)\phi = \xi\phi$. Hence $\alpha_H(h) = \xi$ because ϕ maps $G^{\wedge}$ onto $H^{\wedge}$ (H was assumed to be dually embedded in G). ∎

(14.7) LEMMA. Let H be a closed subgroup of an abelian topological group G. If α_G is continuous, so is $\alpha_{G/H}$.

Proof. Let $\psi : G \to G/H$ be the canonical projection. A direct verification shows that the diagram

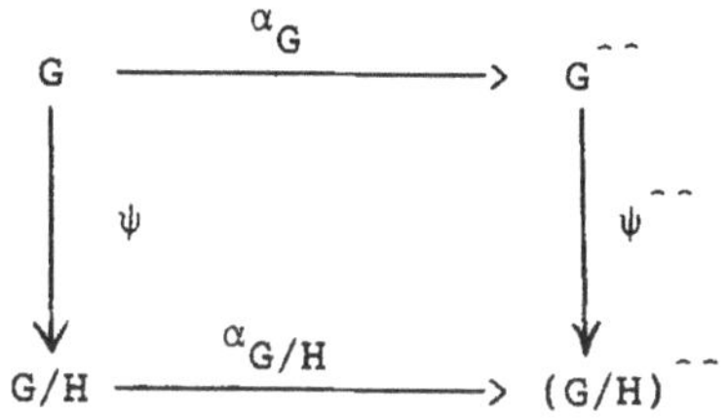

is commutative. ∎

Let H be a closed subgroup of an abelian topological group G. The canonical homomorphisms $G^{\wedge}/H^{\circ} \to H^{\wedge}$ and $(G/H)^{\wedge} \to H^{\circ}$, defined in the obvious way, will be denoted by ϕ_H and ϕ^H, respectively. Observe that ϕ_H is a continuous injection; it is a surjection if and only if H is dually embedded in G. The mapping ϕ^H is a continuous isomorphism of $(G/H)^{\wedge}$ onto H°.

(14.8) LEMMA. Let H be a dually closed and dually embedded subgroup of a Hausdorff locally quasi-convex group G. If $\alpha_G(G) = G^{\wedge\wedge}$ and the group $G^{\wedge}/H^{\circ}$ is locally quasi-convex, then $\phi_H : G^{\wedge}/H^{\circ} \to H^{\wedge}$ is a topological isomorphism.

Proof. The Hausdorff locally quasi-convex group G admits sufficiently many continuous characters, which means that α_G is injective. From (14.3) it follows that α_G is open, i.e. $\alpha_G^{-1} : G^{\wedge\wedge} \to G$ is continuous. By (14.2), we have $\alpha_G(H) = H^{\circ\circ}$. Therefore, the homomorphism $\beta : H^{\circ\circ} \to H$, defined as the restriction of α_G^{-1} to $H^{\circ\circ}$, is continuous. Let $Q = G^{\wedge}/H^{\circ}$ and let $\gamma : H^{\wedge} \to Q^{\wedge\wedge}$ be the homomorphism dual to $\beta\phi^{H^{\circ}} : Q^{\wedge} \to H$. Since β and $\phi^{H^{\circ}}$ are both surjective, γ is injective. A direct verification shows that $\gamma\phi_H = \alpha_Q$. Let $Q^{\vee} = \alpha_Q(Q)$; since H is dually embedded, ϕ_H maps Q onto $H^{\wedge}$, so that we obtain the following commutative diagram

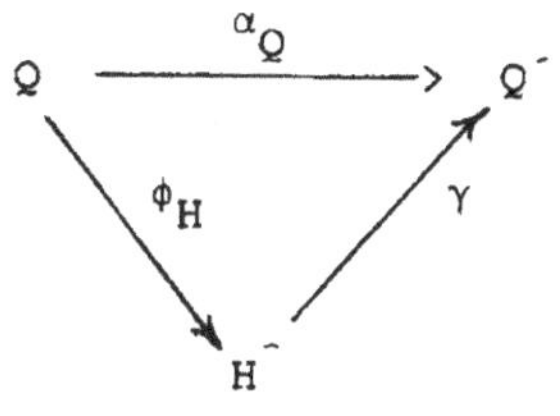

where γ is a continuous isomorphism. To complete the proof, it remains to observe that, due to (14.3), the mapping $\alpha_Q : Q \to Q'$ is open. ∎

Let f be a continuous mapping of a topological space X onto a topological space Y. We say that f is compact-covering if to each compact subset A of Y there corresponds a compact subset B of X such that $A \subset f(B)$.

(14.9) PROPOSITION. An open mapping of a Čech-complete space X onto a topological space Y is compact-covering.

This fact was proved by Arkhangelskiĭ [2].

(14.10) COROLLARY. Let H be a closed subgroup of a Čech-complete abelian group G. Then $\phi^H : (G/H)^{\wedge} \to H^o$ is a topological isomorphism. ∎

(14.11) PROPOSITION. Let $\{G_i\}_{i\in I}$ be a family of abelian topological groups. There are canonical topological isomorphisms

$$(\prod_{i\in I} G_i)^{\wedge} \sim \sum_{i\in I}{}^* G_i^{\wedge} \qquad \text{and} \qquad (\sum_{i\in I}{}^* G_i)^{\wedge} \sim \prod_{i\in I} G_i^{\wedge}.$$

If all groups G_i are reflexive, so are $\prod_{i\in I} G_i$ and $\sum_{i\in I}{}^* G_i$.

Proof. To simplify the notation, we shall write Π_i and Σ_i^* instead of $\prod_{i\in I}$ and $\sum_{i\in I}{}^*$, respectively. Let $\chi \in (\Pi_i G_i)^{\wedge}$. For each $i \in I$, let χ_i denote the restriction of χ to G_i (we identify G_i with the corresponding subgroup of $\Pi_i G_i$). It is clear that $\chi_i = 0$ for all but finitely many indices i. Consider the canonical mapping

$$\phi : (\Pi_i G_i)^{\wedge} \to \Sigma_i^* G_i^{\wedge}$$

given by $\phi(\chi) = (\chi_i)_{i\in I}$. It is clear that ϕ is an algebraic isomorphism. To prove that it is continuous, choose any $U \in N_o(\Sigma_i^* G_i^{\wedge})$. We may assume that $U = \Sigma_i^* U_i$ where $U_i \in N_o(G_i^{\wedge})$ for every i. For each $i \in I$, we can find some compact subset K_i of G_i such that $K_i^o + K_i^o \subset U_i$. Then $K : \Pi_i K_i$ is a compact subset of $\Pi_i G_i$ and it is enough to show that $\phi(K^o) \subset U$.

So, take any $\chi \in K^o$. We may assume that all sets K_i are symmetric. Then it is not hard to see that

$$\Sigma_i |\chi(K_i)| = |\chi(K)| \leq \frac{1}{4}.$$

Write $c_i = |\chi(K_i)|$ for every i. From the condition $K_i^o + K_i^o \subset U$ it follows easily that

$$\chi_i/U_i \leq \frac{1}{2}[\frac{1}{4c_i}]^{-1}$$

where $[x]$ is the integer part of x. Since $\Sigma_i c_i \leq \frac{1}{4}$, after easy calculations we get $\Sigma_i(\chi_i/U_i) < 1$, which means that $\phi(\chi) = (\chi_i)_{i\in I} \in \Sigma_i^* U_i = U$.

To prove that ϕ is open, choose any $W \in N_o((\Pi_i G_i)^\wedge)$. We may assume that $W = K^o$ for some compact subset K of $\Pi_i G_i$. Next, we assume that $K = \Pi_i K_i$ where K_i is a compact subset of G_i for every i. Then $K_i^o \in N_o(G_i^\wedge)$ for every i, so that $\Sigma_i^* K_i^o \in N_o(\Sigma_i^* G_i^\wedge)$ and it remains to show that $\Sigma_i^* K_i^o \subset \phi(W)$.

Take any $\chi \in (\Pi_i G_i)^\wedge$ with $\phi(\chi) = (\chi_i)_{i\in I} \in \Sigma_i^* K_i^o$. We have to prove that $|\chi(K)| \leq \frac{1}{4}$. It is not hard to verify that $\chi_i/K_i^o \geq 4|\chi_i(K_i)|$ for every i. Then

$$|\chi(K)| \leq \Sigma_i |\chi_i(K_i)| \leq \frac{1}{4}\Sigma_i(\chi_i/K_i^o) < \frac{1}{4}.$$

The proof that $(\Sigma_i^* G_i)^\wedge$ is canonically topologically isomorphic to $\Pi_i G_i^\wedge$ is similar, and even a little bit simpler (use (1.17)).

The last assertion of (14.11) is a consequence of the following canonical isomorphisms:

$$(\Pi_i G_i)^{\wedge\wedge} \sim (\Sigma_i^* G_i^\wedge)^\wedge \sim (\Pi_i G_i^{\wedge\wedge}) \sim \Pi_i G_i,$$

$$(\Sigma_i^* G_i)^{\wedge\wedge} \sim (\Pi_i G_i^\wedge)^\wedge \sim (\Sigma_i^* G_i^{\wedge\wedge}) \sim \Sigma_i^* G_i. \quad \blacksquare$$

(14.12) PROPOSITION. Let $G = \sum_{i\in I} (G_i : H_i)$ be a reduced product of abelian topological groups. Suppose that H_i is dually closed in G_i for almost all i. Then $G^\wedge$ is canonically topologically isomorphic to the reduced product $\sum_{i\in I} (G_i^\wedge : H_i^\circ)$. Consequently, if all groups G_i are reflexive, then G is reflexive, too.

We omit the proof because it is very similar to that of (18.2) below. For details, we refer the reader to [60]. See also [38], (23.33).

Let $G = \sum_{i\in I} (G_i : H_i)$ be a reduced product of abelian topological groups. Let H be a subgroup of G consisting of all sequences (g_i) such that $g_i \in H_i$ for all i. Then H may be identified with the product $\prod_{i\in I} H_i$ with the usual product topology and G/H may be identified with the direct sum $\sum_{i\in I} (G_i/H_i)$. We may also identify H° with the subgroup $\prod_{i\in I} H_i^\circ$ of $\sum_{i\in I} (G_i^\wedge : H_i^\circ)$. It is clear that H is dually closed (resp. dually embedded) in G if and only if H_i is dually closed (resp. dually embedded) in G_i for every i. Similarly, G/H is locally quasi-convex if and only if G_i/H_i is locally quasi-convex for every i. The mapping $\phi_H : G^\wedge/H^\circ \to H^\wedge$ (resp. $\phi^H : (G/H)^\wedge \to H^\circ$) is a topological isomorphism if and only if $\phi_{H_i} : G_i^\wedge/H_i^\circ \to H_i^\wedge$ (resp. $\phi^{H_i} : (G_i/H_i)^\wedge \to H_i^\circ$) is a topological isomorphism for every i. Detailed proofs of these facts can be found in [60].

(14.13) NOTES. The material of this section is taken mainly from [8]. Proposition (14.11) is known as Kaplan's duality theorem and was proved by Kaplan [49]. For countable products, it was obtained independently by Vilenkin (see [97], n° 4). Proposition (14.12) is taken from [60]. It had been obtained earlier by Vilenkin [97], Theorem 7. In the case when G_i are LCA groups and H_i are open compact subgroups, (14.12) was proved independently by Braconnier [25] and Vilenkin [96].

15. Locally convex vector groups

(15.1) PROPOSITION. A locally convex space E satisfies the condition $\alpha_E(E) = E^{\wedge\wedge}$ if and only if closed convex hulls of compact subsets of E are weakly compact.

Proof. By (2.3), we have a topological isomorphism $\rho_E : E_c^* \to E^\wedge$ and an algebraical isomorphism $\rho_{E_c^*} : (E_c^*)^* \to (E_c^*)^\wedge$. Let $\beta : E \to (E_c^*)^*$ be the canonical embedding. We have the following commutative diagram:

$$\begin{array}{ccccc} E & & \xrightarrow{\quad id \quad} & & E \\ \Big\downarrow \alpha_E & & & & \Big\downarrow \beta \\ E^{\wedge\wedge} & \xrightarrow{\rho_E^{\wedge}} & (E_c^*)^{\wedge} & \xleftarrow{\rho_{E_c^*}} & (E_c^*)^* \end{array}$$

So, the condition $\alpha_E(E) = E^{\wedge\wedge}$ is equivalent to

(1) $\beta(E) = (E_c^*)^*$.

Suppose first that closed convex hulls of compact subsets of E are weakly compact. We have to prove (1). Let $h \in (E_c^*)^*$. There is a symmetric compact subset A of E such that, by denoting

$$A^0 = \{f \in E : |f(u)| \leq 1 \text{ for all } u \in A\},$$

one has

(2) $|h(f)| \leq 1$ for all $f \in A^0$.

Let $\mathfrak{m}$ be the family of all finite dimensional subspaces of E. Fix an arbitrary $M \in \mathfrak{m}$. It is clear that we can find some $v \in E$ with

(3) $f(v) = h(f)$ for all $f \in M$.

Let us denote

$$M^{\perp} = \{u \in E : f(u) = 0 \text{ for all } f \in M\}.$$

We shall prove that

(4) $v \in M^{\perp} + \overline{\text{conv } A}$.

Suppose the contrary. Since $M^{\perp}$ is weakly closed and, by our assumption, $\overline{\text{conv } A}$ is weakly compact, it follows that $M^{\perp} + \overline{\text{conv } A}$ is weakly closed in E. So, by the Hahn-Banach theorem, there is some $f_o \in E^*$ with $|f_o(v)| > 1$ and

(5) $|f_o(u)| \leq 1$ for all $u \in M^{\perp} + \overline{\text{conv } A}$.

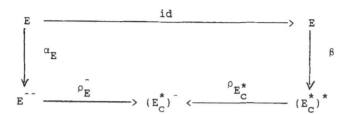

From (5) it follows that $f_o \in M \cap A^0$. Hence, by (3) and (2), we obtain $|f_o(v)| = |h(f_o)| \leq 1$, which is a contradiction.

Now, (4) implies that there is some $w \in \overline{\text{conv } A}$ with $w - v \in M^{\perp}$. Then, by (3), for each $f \in M$, one has

$$f(w) = f(v) + f(w - v) = h(f).$$

We have thus shown that, for each $M \in \mathfrak{M}$, the set

$$P_M := \overline{\text{conv } A} \cap \{u \in E : h(f) = f(u) \text{ for all } f \in M\}$$

is non-empty. Since $\overline{\text{conv } A}$ is weakly compact, the intersection of all P_m´s contains some vector v. Then $f(v) = h(f)$ for all $f \in E$, i.e. $\beta(v) = h$. This proves (1).

Conversely, suppose that (1) is satisfied. Choose an arbitrary compact subset A of E and denote

$$A^0 = \{f \in E^* : |f(u)| \leq 1 \text{ for all } u \in A\},$$

$$B = \{u \in E : |f(u)| \leq 1 \text{ for all } f \in A^0\}.$$

Naturally, B is a closed, convex subset of E containing A, so that it remains to show that B is weakly compact. Since $A^0 \in N_o(E_c^*)$, the Banach-Alaoglu theorem implies that

$$A^{00} := \{h \in (E_c^*)^* : |h(f)| \leq 1 \text{ for all } f \in A^0\}$$

is compact in the weak* topology on $(E_c^*)^*$. Now, (1) implies that $\beta(B) = A^{00}$ and it remains to observe that β is a topological isomorphism of the space E endowed with its weak topology onto the space $(E_c^*)^*$ endowed with its weak* topology. ∎

(15.2) PROPOSITION. Every metrizable and complete locally convex vector space is a reflexive group.

"Vector space" may be replaced here by "vector group"; see (15.7).

Proof. Let E be a metrizable and complete locally convex space. Then closed convex hulls of compact subsets of E are compact, hence weakly compact, and (15.1) says that $\alpha_E(E) = E^{\wedge\wedge}$. That α_E is open and continuous follows from (14.3) and (14.4). ∎

(15.3) PROPOSITION. Every reflexive locally convex space is a reflexive group.

Proof. Let E be a reflexive locally convex space. Then closed bounded subsets of E are weakly compact ([80], Ch. IV, Theorem 5.5), and (15.1) implies that $\alpha_E(E) = E^{\wedge\wedge}$. From (2.4) and (14.3) we infer that α_E is open. We shall prove that it is continuous.

In view of (2.3), it is enough to show that compact subsets of E_c^* are equicontinuous. So, let A be a compact subset of E_c^*. Since E is reflexive, it suffices to show that A is a bounded subset of E_b^*. Suppose the contrary; then there is some $f \in (E_b^*)^*$ with $\sup\{|f(u)|: u \in A\} = \infty$. However, since E is reflexive, we have $f \in (E_c^*)^*$ and then $\sup\{|f(u)| : u \in A\} < \infty$ because A was compact in E_c^*. The contradiction obtained completes the proof. ∎

(15.4) REMARK. Since there exist non-reflexive Banach spaces, from (15.2) and (15.3) it follows that, for locally convex spaces, reflexivity is an essentially stronger property than "group reflexivity". See also (18.2). One more notion of reflexivity can be obtained by considering the so-called structure of continuous convergence on topological groups. H.-P. Butzmann [27] proved that a locally convex space is reflexive in the sense of continuous convergence if and only if it is complete.

(15.5) REMARK. Kye [57] proved that a locally convex space E is a reflexive group if and only if closed convex hulls of compact subsets of E are weakly compact and each closed convex balanced set which is a neighbourhood of zero in the k-topology is a neighbourhood of zero in the original topology. The first condition is satisfied if and only if $\alpha_E(E) = E^{\wedge\wedge}$ (see (15.1)). The second one is equivalent to the continuity of α_E.

(15.6) LEMMA. Let E be a locally convex space, L a vector space endowed with the discrete topology and M a linear subspace of $E \times L$. For each $u \in (E \times L) \setminus \overline{M}$, there exists some $\chi \in (E \times L)^{\wedge}$ with $\chi_{|M} = 0$ and $\chi(u) \neq 0$.

Proof. Choose any $u \in (E \times L) \setminus \overline{M}$. Let ψ_E and ψ_L be canonical projections of $E \times L$ onto E and L, respectively. Suppose first that $\psi_L(u) \notin \psi_L(M)$. Since L is discrete, there is some $\kappa \in L^{\wedge}$

with $\kappa(\psi_L(M)) = \{0\}$ and $\kappa(\psi_L(u)) \neq 0$; then we may take $\chi = \kappa\psi_L$.

Next, suppose that $\psi_L(u) \in \psi_L(M)$. Then $\psi_L(u) = \psi_L(w)$ for some $w \in M$. Let $N = \psi_E(M \cap (E \times \{0\}))$. Since $u \notin \overline{M}$, it follows that $\psi_E(u - w) \notin \overline{N}$. So, there is a continuous linear functional f on E such that $f_{|N} = 0$ and $f\psi_E(u - w) \notin Z$. Then $\rho f \in E^\wedge$, $\rho f_{|N} = 0$ and $\rho f\psi_E(u - w) \neq 0$. It is clear that the formula

$$\kappa(\psi_L(v)) = \rho f\psi_E(v) \qquad (v \in M)$$

defines a character κ of $\psi_L(M)$. Then the formula

$$\xi(v) = \rho f\psi_E(v) - \kappa\psi_L(v) \qquad (v \in E \times \psi_L(M))$$

defines a continuous character ξ of the group $E \times \psi_L(M)$. Naturally, we may extend ξ to some $\chi \in (E \times L)^\wedge$. Then $\chi_{|M} = \xi_{|M} = 0$ and

$$\chi(u) = \xi(u) = \rho f\psi_E(u) - \kappa\psi_L(u) = \rho f\psi_E(u) - \kappa\psi_L(w)$$

$$= \rho f\psi_E(u) - \rho f\psi_E(w) = \rho f\psi_E(u - w) \neq 0. \quad \blacksquare$$

(15.7) THEOREM. Every metrizable and complete locally convex vector group is reflexive.

Proof. Let F be a metrizable and complete locally convex vector group. Let $\{U_n\}_{n=1}^{\infty}$ be a base at zero consisting of symmetric convex sets. Fix an arbitrary $n = 1,2,...$ and denote $M_n = \operatorname{span} U_n$. We can find a linear subspace L_n of F such that $F = M_n \oplus L_n$ algebraically. Since M_n is an open subgroup of F, it follows that L_n is discete and $F = M_n \oplus L_n$ topologically. Hence the canonical projections $\pi_n^M : F \to M_n$ and $\pi_n^L : F \to L_n$ are continuous. Let p_n be the Minkowski functional of U_n in M_n and let E_n be the quotient space $M_n/p_n^{-1}(0)$ endowed with the canonical norm. Let then $\psi_n : M_n \to E_n$ be the canonical projection and let F_n denote the group F endowed with the topology induced by the homomorphism

$$\phi_n = \psi_n\pi_n^M \times \pi_n^L : F \to E_n \times L_n.$$

Denote $G = \prod_{n=1}^{\infty} (\tilde{E}_n \times L_n)$ and consider the linear mapping $\phi =$

$(\phi_n)_{n=1}^{\infty} : F \to G$. It is clear that F may be identified, as a topological group, with the limit of the inverse sequence $id : F_{n+1} \to F_n$. Therefore ϕ is a topological isomorphism between F and $\phi(F)$. From (1.8), (15.2) and (14.11) we see that G is reflexive. In view of (14.3), (14.4) and (14.6), to complete the proof, we only have to show that $\phi(F)$ is dually embedded and dually closed in G.

So, take any $\chi \in \phi(F)^{\wedge}$. Then $\chi\phi \in F^{\wedge}$ and, consequently, $\chi\phi \in F_n^{\wedge}$ for some n because $F^{\wedge} = \bigcup_{n=1}^{\infty} F_n^{\wedge}$. Since the topology of F_n is induced by ϕ_n, we can write $\chi\phi = \kappa\phi_n$ for some $\kappa \in (E_n \times L_n)^{\wedge}$. Let $\tilde{\kappa} \in (\tilde{E}_n \times L_n)^{\wedge}$ be the natural extension of κ and let $\sigma_n : G \to \tilde{E}_n \times L_n$ be the canonical projection. An easy verification shows that $\tilde{\kappa}\sigma_n$ is an extension of χ. Thus $\phi(F)$ is dually embedded in G.

To prove that $\phi(F)$ is dually closed, choose any $g \in G \setminus \phi(F)$. Since F is complete, $\phi(F)$ is a closed subgroup of G. Therefore we can find some $U \in N_0(G)$ with $(g + U) \cap \phi(F) = \emptyset$. In other words, there is a finite subset $I \subset \{1,2,\dots\}$ such that if σ_I is the canonical projection of G onto $G_I := \prod_{n\in I} (\tilde{E}_n \times L_n)$, then $\sigma_I(g)$ does not belong to the closure of $\sigma_I(\phi(F))$ in G_I. Hence, by (15.6), there is some $\chi \in G_I^{\wedge}$ with $\chi(\sigma_I(\phi(F))) = \{0\}$ and $\chi(\sigma_I(g)) \neq 0$. Then $\chi\sigma_I$ is a continuous character of G taking $\phi(F)$, but not g, to zero. ∎

(15.8) REMARK. The above argument shows that if F is a complete locally convex vector group not necessarily metrizable, then $\alpha_F(F) = F^{\wedge\wedge}$. From (9.4), (8.5) and (14.3) it follows that α_F is open. The following example shows that α_F need not be continuous (see also (17.6)).

(15.9) EXAMPLE. Let E be an infinite dimensional normed space endowed with its weak topology. Clearly, there is a convex infinite dimensional subset K of E^* compact in the norm topology. Since weakly compact subsets of E are bounded, K is compact in E_c^* but, being infinite dimensional, it cannot be equicontinuous. Thus the evaluation map $E \to (E_c^*)_c^*$ is not continuous and, by (2.3), neither is α_E.

(15.10) NOTES. Propositions (15.1) and (15.2) are taken from [57] (cf. (15.5)). Proposition (15.1) is an immediate consequence of the result of Arens [1], Theorem 2, which asserts that if E is a locally convex space and τ is a locally convex topology on E^* not weaker than the topology of pointwise convergence, then the evaluation map $E \to (E^*_\tau)^*$ is onto if and only if τ is not stronger than the topology of uniform convergence on convex weakly compact subsets of E. Arens's paper contains a very detailed study of "reflexive" topologies on E^* and E^{**}.

For Banach spaces, (15.2) appears in [83]. Also (15.3) is taken from [83]. Lemma (15.6) and Theorem (15.7) are new. Example (15.9) is due to G.W. Mackey and was given in [1].

16. Nuclearity of dual groups

The aim of this section is to prove the following fact:

(16.1) THEOREM. Let G be a nuclear group. If G is metrizable, then $G^\wedge_{pc}$ is a nuclear group, too.

(16.2) REMARK. The assumption of metrizability is essential. Let E be an uncountable product of real lines. Then E is a nuclear locally convex space, hence a nuclear group. By (14.11), we may identify $E^\wedge_{pc}$ with an uncountable locally convex direct sum of real lines. As a locally convex space, an uncountable direct sum of real lines is not nuclear. So, by virtue of (8.9), the group $E^\wedge_{pc}$ cannot be nuclear.

(16.3) LEMMA. Let F be a metrizable nuclear vector group and let P be a precompact subset of F. Given arbitrary $c > 0$ and $m = 1, 2, \ldots$, one can find a vector space E, two symmetric and convex subsets X, Y of E with

(1) $$d_k(X,Y) \leq ck^{-m} \qquad (k = 1,2,\ldots),$$

a subgroup K of E and a homomorphism $\phi : K \to (F)$, such that $\phi(K \cap Y)$ is precompact and $P \subset \phi(K \cap X)$.

Proof. By (9.3), one can find a base $\{U_n\}_{n=1}^{\infty}$ at zero in F, consisting of symmetric, convex sets such that $U_1 \supset U_2 \supset \ldots$ and

(2) $$d_k(U_{n+1},U_n) < c2^{-2mn-1}\, k^{-m} \qquad (k,n = 1,2,\ldots).$$

Let I be a set of indices. By R_o^I we denote the corresponding direct sum of real lines, i.e. the subspace of the product R^I consisting of all functions $f : I \to R$ such that $f(i) = 0$ for almost all $i \in I$. By Z_o^I we denote the subgroup of R_o^I consisting of integer-valued functions.

Fix an arbitrary $n = 1,2,\ldots$ and let $M_n = \operatorname{span} U_n$. There is a linear subspace L_n of F such that $F = M_n \oplus L_n$ (that is, the vector space F is the algebraical direct sum of its vector subspaces M_n and L_n; since M_n is an open subset of F, it follows that the topological group F is the topological direct sum of its subgroups M_n and L_n). Since every abelian group is a quotient of a free one, we may identify L_n with a quotient group of $Z_o^{I_n}$ for some set I_n; let $\psi_n : Z_o^{I_n} \to L_n$ be the canonical projection.

There is a finite subset A_n of F with $P \subset U_{n+1} + A_n$ because P is precompact. Since $F = M_n \oplus L_n$, we have $A_n \subset B_n + C_n$ for some finite subsets B_n of M_n and C_n of L_n; we may assume that B_n and C_n are symmetric. Next, there is a symmetric and finite subset D_n of $Z_o^{I_n}$ with $\psi_n(D_n) = C_n$. Thus

(3) $$P \subset U_{n+1} + B_n + \psi_n(D_n).$$

It is clear that we can find a symmetric and finite subset B_n' of M_n such that

(4) $$d_k(\operatorname{conv} B_n, \operatorname{conv} B_n') < c2^{-2mn-1}k^{-m} \qquad (k = 1,2,\ldots).$$

Similarly, there is a finite and symmetric subset D_n' of $R_n^{I_n}$ such that

(5) $$d_k(\operatorname{conv} D_n, \operatorname{conv} D_n') < c2^{-2mn}k^{-m} \qquad (k = 1,2,\ldots).$$

Let us define

$$X_{2n-1} = U_{n+1} + \operatorname{conv} B_n, \qquad Y_{2n-1} = U_n + \operatorname{conv} B_n',$$

$$X_{2n} = \operatorname{conv} D_n, \qquad Y_{2n} = \operatorname{conv} D_n'.$$

Now, take any $k = 1,2,\ldots$. Applying (2.6) (a) and then (2) and (4), we get

(6) $$d_{2k-1}(X_{2n-1},Y_{2n-1}) = d_{2k-1}(U_{n+1} + \text{conv } B_n,\ U_n + \text{conv } B_n')$$
$$\le d_k(U_{n+1},\ U_n + \text{conv } B_n') + d_k(\text{conv } B_n,\ U_n + \text{conv } B_n')$$
$$\le d_k(U_{n+1},U_n) + d_k(\text{conv } B_n,\ \text{conv } B_n')$$
$$< c2^{-2mn-1}\, k^{-m} + c2^{-2mn-1}\, k^{-m} = c2^{-2mn}\, k^{-m}.$$

Hence

(7) $$d_{2k-1}(X_{2n-1},Y_{2n-1}) < c2^{-m(2n-1)}(2k-1)^{-m}.$$

From (6) we also obtain

(8) $$d_{2k}(X_{2n-1},Y_{2n-1}) \le d_{2k-1}(X_{2n-1},Y_{2n-1})$$
$$< c2^{-2mn}\, k^{-m} = c2^{-m(2n-1)}(2k)^{-m}.$$

From (7) and (8) we derive

(9) $$d_k(X_{2n-1},Y_{2n-1}) \le c2^{-m(2n-1)}\, k^{-m} \qquad (k,n = 1,2,\ldots).$$

Let us now consider the vector space

$$E = \prod_{n=1}^{\infty} (M_n \times R_o^{I_n}).$$

It consists of sequences $(u_n,f_n)_{n=1}^{\infty}$ where $u_n \in M_n$ and $f_n \in R_o^{I_n}$ for every n. Let K be the subgroup of E consisting of all sequences $(u_n,f_n)_{n=1}^{\infty}$ such that $f_n \in Z_o^{I_n}$ and $u_n + \psi_n(f_n) = u_1 + \psi_1(f_1)$ for every n. Next, let $\phi : K \to F$ be the homomorphism given by the formula $\phi((u_n,f_n)_{n=1}^{\infty}) = u_1 + \psi_1(f_1)$. Finally, set

$$X = \{(u_n,f_n)_{n=1}^{\infty} \in E : u_n \in X_{2n-1} \text{ and } f_n \in X_{2n} \text{ for all } n\},$$

$$Y = \{(u_n,f_n)_{n=1}^{\infty} \in E : u_n \in Y_{2n-1} \text{ and } f_n \in Y_{2n} \text{ for all } n\}.$$

We shall prove that the E,X,Y,K and ϕ thus defined have the desired properties.

Inequality (1) follows from (5), (9) and (2.7). To prove the inclusion $P \subset \phi(K \cap X)$, take any $g \in P$. By (3), to each n there correspond some $u_n \in U_{n+1}$, $b_n \in B_n$ and $d_n \in D_n$, such that $g = u_n + b_n + \psi_n(d_n)$. Then the sequence $(u_n + b_n, d_n)_{n=1}^{\infty}$ belongs to $K \cap X$, and

$$\phi((u_n + b_n, d_n)_{n=1}^{\infty}) = g.$$

It remains to prove that $\phi(K \cap Y)$ is precompact. Let us take any $n = 1,2,\ldots$. The subspace $M_n = \operatorname{span} U_n$ is absorbed by U_n. Since B_n' is finite, we can find another finite subset B_n'' of M_n such that $\operatorname{conv} B_n' \subset B_n'' + U_n$. Then

$$(10) \qquad Y_{2n-1} \subset 2U_n + B_n''.$$

It is clear that the set $D_n'' = Z_o^{I_n} \cap \operatorname{conv} D_n'$ is finite.

Take an arbitrary $g \in \phi(K \cap Y)$. We have $g = \phi((u_n, f_n)_{n=1}^{\infty})$ for some sequence $(u_n, f_n)_{n=1}^{\infty} \in K \cap Y$. Then $u_n \in Y_{2n-1}$ and $f_n \in Y_{2n} \cap Z_o^{I_n}$ for every n. Applying the definition of ϕ and then (10), for every n, we get

$$g = u_n + \psi_n(f_n) \in Y_{2n-1} + \psi_n(Y_{2n} \cap Z_o^{I_n}) \subset 2U_n + B_n'' + \psi_n(D_n'').$$

Since $g \in \phi(K \cap Y)$ was arbitrary, we derive

$$\phi(K \cap Y) \subset 2U_n + B_n'' + \psi_n(D_n'') \qquad (n = 1,2,\ldots).$$

This proves that $\phi(K \cap Y)$ is precompact becasue $\{2U_n\}_{n=1}^{\infty}$ is a base at zero in F. ∎

(16.4) LEMMA. Let K be a subgroup of a vector space E and let X,Y be two symmetric, convex subsets of E such that

$$(1) \qquad d_k(X,Y) \le ck^{-m} \qquad (k = 1,2,\ldots)$$

where $c > 0$ and $m = 5,6,\ldots$ are some fixed numbers. Let ϕ be a homomorphism of K into a topological group G. Then there exist a vector space E', a subgroup K' of E', two symmetric, convex subsets X',Y' of E' and a homomorphism $\phi' : K' \to G^{\hat{}}$, such that

(2) $$[\phi(K \cap Y)]^{o} \subset \phi'(K' \cap X'),$$

(3) $$\phi'(K' \cap Y') \subset [\phi(K \cap X)]^{o},$$

(4) $$d_k(X',Y') \leq \leq c\gamma_m \, k^{-m+5} \qquad (k = 1,2,\ldots)$$

where γ_m is some universal constant depending on m only.

Proof. Set $F = \operatorname{span} X$. From (1) and (2.14) it follows that there are pre-Hilbert seminorms p,q on F with $X \subset B_p \prec B_q \subset Y$ and

$$d_k(B_p,B_q) \leq cc_m \, k^{-m+2} \qquad (k = 1,2,\ldots)$$

where c_m is some universal constant depending on m only. By (2.15), there is another pre-Hilbert seminorm r on E such that

(5) $$d_k(B_p,B_r) \leq 10^2 cc_m \, k^{-m+5} \qquad (k = 1,2,\ldots),$$

(6) $$d_k(B_r,B_q) \leq 10^{-2} \, k^{-3} \qquad (k = 1,2,\ldots).$$

From (6) we get

(7) $$5 \sum_{k=1}^{\infty} k d_k(B_r,B_q) \leq 5\cdot 10^{-2} \sum_{k=1}^{\infty} k^{-2} < \tfrac{1}{4}.$$

We may identify $G^{\hat{}}$ with a subgroup of T^G. Let $\rho_G : R^G \to T^G$ be the canonical projection. Set $E' = R^G$ and $K' = \rho_G^{-1}(G^{\hat{}})$; let ϕ' be the restriction of ρ_G to K'.

For each $g \in \phi(K \cap B_p)$, choose some $w_g \in K \cap B_p$ with $\phi(w_g) = g$. Then the formula

$$(\Phi f)(g) = \begin{cases} f(w_g) & \text{if} \quad g \in \phi(K \cap B_p) \\ 0 & \text{if} \quad g \notin \phi(K \cap B_p) \end{cases}$$

defines a linear operator $\Phi : F^{\#} \to R^G$. Let us denote

$$B_r^o = \{f \in F^\# : |f(u)| \le \tfrac{1}{4} \quad \text{for all} \quad u \in B_r\},$$

$$B_p^o = \{f \in F^\# : |f(u)| \le \tfrac{1}{4} \quad \text{for all} \quad u \in B_p\},$$

$$L = \{\zeta \in R^G : \zeta(g) = 0 \quad \text{for all} \quad g \in \phi(K \cap B_p)\}.$$

Set $X' = \Phi(B_r^o) + L$ and $Y' = \Phi(B_p^o) + L$. Since L is a linear subspace of R^G, we have $d_1(L,L) = 0$. Applying (2.6) (a), (2.8)(a) and (2.16), we get

$$d_k(X',Y') \le d_k(\Phi(B_r^o),\Phi(B_p^o) + L) + d_1(L,\Phi(B_r^o) + L)$$
$$\le d_k(\Phi(B_r^o),\Phi(B_p^o)) + d_1(L,L) \le d_k(B_r^o,B_p^o) = d_k(B_p,B_r)$$

for $k = 1,2,\ldots$. Hence, by (5), we obtain (4) with $\gamma_m = 10^2 c_m$.

To prove (3), take any $\xi \in K' \cap Y'$ and $g \in \phi(K \cap X)$. We may write $\xi = \Phi f + \zeta$ for some $f \in B_p^o$ and $\zeta \in L$. Then

$$|\phi'(\xi)(g)| = |\rho_G(\xi)(g)| = |\rho(\xi(g))| \le |\xi(g)|$$
$$\le |(\Phi f)(g)| + |\zeta(g)| = |f(w_g)| \le \tfrac{1}{4}.$$

Thus $\phi'(\xi) \in (\phi(K \cap X))^o$, which proves (3).

It remains to prove (2). So, take any $\chi \in (\phi(K \cap Y))^o$. Let $H = K \cap F$ and $\kappa = \chi\phi_{|H}$. Then $|\kappa(H \cap B_q)| \le \frac{1}{4}$. By (7) and (8.1), there is some $f \in B_r^o$ with $\rho f_{|H} = \kappa$. Then the formula

$$\xi(g) = \begin{cases} \chi(g) & \text{if} \quad g \notin \phi(H \cap B_p) \\ f(w_g) & \text{if} \quad g \in \phi(H \cap B_p) \end{cases}$$

defines some function $\xi \in R^G$. It is easy to verify that $\rho_G(\xi) = \chi$ and $\xi - \Phi f \in L$. Hence $\xi \in K' \cap X'$ and $\phi'(\xi) = \chi$. ▪

(16.5) LEMMA. Let F be a nuclear vector group. If F is metrizable, then $\hat{F}_{pc}$ is a nuclear group.

Proof. Choose arbitrary $U \in \mathcal{N}_0(F^{\wedge}_{pc})$, $c > 0$ and $m = 1,2,\ldots,$ There is a precompact subset P of F with $P^0 \subset U$. According to (16.3), there exist a vector space E, two symmetric and convex subsets X,Y of E with

$$d_k(X,Y) \leq ck^{-m-5} \qquad (k = 1,2,\ldots),$$

a subgroup K of E and a homomorphism $\phi : K \to F$, such that $\phi(K \cap Y)$ is precompact and $P \subset \phi(K \cap X)$. Next, by (16.4), there exist a vector space E', a subgroup K' of E', two symmetric convex subsets X',Y' of E' and a homomorphism $\phi' : K' \to F$, such that

$$[\phi(K \cap Y)]^0 \subset \phi(K' \cap X'), \qquad \phi'(K' \cap Y') \subset [\phi(K \cap X)]^0,$$

$$d_k(X',Y') \leq c\gamma_m k^{-m} \qquad (k = 1,2,\ldots)$$

where γ_m is a universal constant depending on m only. Now, it remains to observe that

$$\phi'(K' \cap Y') \subset [\phi(K \cap X)]^0 \subset P^0 \subset U,$$

and that $\phi'(K' \cap X') \in \mathcal{N}_0(F^{\wedge}_{pc})$ because $[\phi(K \cap Y)]^0 \subset \phi'(K' \cap X')$ and $\phi(K \cap Y)$ is precompact. ■

Proof of (16.1). Every compact subset of $\tilde{G}$ is contained in the closure of some precompact subset of G. Therefore we may identify $G^{\wedge}_{pc}$ with $(\tilde{G})^{\wedge}_c$. So, we have to prove that $G^{\wedge} = G^{\wedge}_c$ is nuclear provided that G is complete.

According to (9.7), there are a metrizable and complete nuclear vector group F, a closed subgroup H of F and a closed subgroup Q of H, such that G is topologically isomorphic to H/Q. From (9.4), (8.3) and (8.6) we infer that H is dually embedded and dually closed in F. From (16.5) and (7.5) it follows that $F^{\wedge}/H^0$ is a nuclear group. Hence, by (8.5), it is locally quasi-convex. Therefore, by (15.7) and (14.8), the mapping $\phi_H : F^{\wedge}/H^0 \to H^{\wedge}$ is a topological isomorphism. Thus $H^{\wedge}$ is a nuclear group. Applying (7.5), we see that Q^0_H is nuclear, too. To complete the proof, it remains to observe that, due to (14.10), the mapping $\phi^Q : (H/Q)^{\wedge} \to Q^0_H$ is a topological isomorphism. ■

(16.6) NOTE. The material of this section is new.

17. Strong reflexivity

A reflexive group G is called <u>strongly reflexive</u> if every closed subgroup and every Hausdorff quotient group of G and of $G^\wedge$ is reflexive. Strong reflexivity is an essentially stronger property than reflexivity. For instance, every infinite dimensional Banach space is a reflexive, but not strongly reflexive group ((5.3) and (15.2)).

(17.1) PROPOSITION. Let H be a closed subgroup of a strongly reflexive group G. Then

(a) H is dually closed in G;

(b) H is dually embedded in G;

(c) the canonical mappings $\phi_H : G^\wedge/H^\circ \to H^\wedge$ and $\phi^H : (G/H)^\wedge \to H^\circ$ are topological isomorphisms;

(d) H and G/H are strongly reflexive.

Proof. (a) Since G is strongly reflexive, G/H is reflexive. In particular, G/H admits sufficiently many continuous characters, which means that H is dually closed in G.

(b) Since G is reflexive, it follows from (a) and (14.2) that $\alpha_{G|H}$ is a topological isomorphism of H onto $H^{\circ\circ}$. Let $\gamma : H^{\circ\circ} \to H$ be the inverse isomorphism. Choose any $\chi \in H^\wedge$ and consider the continuous isomorphism $\phi^{H^\circ} : (G^\wedge/H^\circ) \to H^{\circ\circ}$. Then $\chi\gamma\phi^{H^\circ} \in (G^\wedge/H^\circ)^{\wedge\wedge}$. By assumption, $G^\wedge/H^\circ$ is reflexive, so that we may write $\chi\gamma\phi^{H^\circ} = \alpha_{G^\wedge/H^\circ}(\xi)$ for some $\xi \in G^\wedge/H^\circ$. Let $\psi : G^\wedge \to G^\wedge/H^\circ$ be the canonical projection. Then $\xi = \psi(\kappa)$ for a certain $\kappa \in G^\wedge$ and a direct verification shows that $\kappa_{|H} = \chi$.

(c) Since G is strongly reflexive, $G^\wedge/H^\circ$ is reflexive, hence locally quasi-convex. It follows from (a), (b) and (14.8) that ϕ_H is a topological isomorphism. Similarly, $\phi_{H^\circ} : G^{\wedge\wedge}/H^{\circ\circ} \to (H^\circ)^\wedge$ is a topological isomorphism. Consider the canonical commutative diagram

$$\begin{array}{ccc} G & \xrightarrow{\alpha_G} & G^{\wedge\wedge} \\ \downarrow & & \downarrow \\ G/H & \xrightarrow{\beta} & G^{\wedge\wedge}/H^{\circ\circ}. \end{array}$$

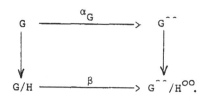

It is clear that β is a topological isomorphism. A direct verification shows that the composition

$$H^{\circ} \xrightarrow{\ \alpha_{H^{\circ}}\ } (H^{\circ})^{\wedge\wedge} \xrightarrow{\ \phi_{H^{\circ}}^{\wedge}\ } (G^{\wedge\wedge}/H^{\circ\circ})^{\wedge} \xrightarrow{\ \beta^{\wedge}\ } (G/H)^{\wedge}$$

is equal to $(\phi^{H})^{-1}$.

(d) Let K be a closed subgroup of H. Then K is a closed subgroup of G, therefore it is reflexive. We shall prove that H/K is reflexive.

Consider the canonical diagram

$$\begin{array}{ccc} H & \xrightarrow{\ i\ } & G \\ \downarrow & & \downarrow \\ H/K & \xrightarrow{\ j\ } & G/K \end{array}$$

It is clear that j is a topological embedding. Since, by (a) and (b), H is dually closed and dually embedded in G, it follows easily that $j(H/K)$ is dually closed and dually embedded in G/K. So, in virtue of (14.6), we have $\alpha_{H/K}(H/K) = (H/K)^{\wedge\wedge}$.

Since G is strongly reflexive, G/K is reflexive, hence locally quasi-convex. Therefore $j(H/K)$ is locally quasi-convex, too (subgroups of locally quasi-convex groups are locally quasi-convex). Consequently, H/K is locally quasi-convex and from (14.3) and (14.7) we infer that $\alpha_{H/K}$ is open and continuous. Thus H/K is reflexive.

Finally, let Q be a closed subgroup of G/H. That Q is reflexive follows from the above and the obvious fact that Q may be identified with a Hausdorff quotient of a closed subgroup of G. The group $(G/H)/Q$ is reflexive because it is a quotient of G. ▪

A nuclear group G is called <u>binuclear</u> if $G^{\wedge}$ is nuclear and α_G maps G onto $G^{\wedge\wedge}$ (we do not assume α_G to be continuous).

(17.2) THEOREM. Let G be a binuclear group. If G is Čech-complete, then it is strongly reflexive.

Proof. That G is reflexive follows from (8.5), (14.3) and (14.4) because Čech-complete space are k-spaces. Every closed subgroup of G

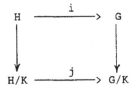

and of $G^\wedge$ is dually closed and dually embedded due to (8.6) and (8.3), respectively. From (7.5) and (8.5) it follows that Hausdorff quotient groups of G and of $G^\wedge$ are locally quasi-convex.

Let H be an arbitrary closed subgroup of G. In view of (14.2), we only need to prove that the four groups H, G/H, H^o and $G^\wedge/H^o$ are reflexive.

As in the proof of (17.1) (c), we see that the canonical mapping $\beta : G/H \to G^{\wedge\wedge}/H^{oo}$ is a topological isomorphism. The group $G^{\wedge\wedge}/H^{oo}$ is locally quasi-convex, being topologically isomorphic to the locally quasi-convex group G/H. So, by (14.8), the mapping

$$\phi_{H^o} : G^{\wedge\wedge}/H^{oo} \to (H^o)^\wedge$$

is a topological isomorphism. Since G is Čech-complete, (14.10) plies that $\phi^H : (G/H)^\wedge \to H^o$ is a topological isomorphism, too. It is not hard to verify that the diagram

$$\begin{array}{ccc} G^{\wedge\wedge}/H^{oo} & \xrightarrow{\ \phi_{H^o}\ } & (H^o)^\wedge \\ \uparrow{\scriptstyle\beta} & & \downarrow{\scriptstyle(\phi^H)^\wedge} \\ G/H & \xrightarrow{\ \alpha_{G/H}\ } & (G/H)^{\wedge\wedge} \end{array}$$

is commutative. Thus $\alpha_{G/H}$ is a topological isomorphism of G/H onto $(G/H)^{\wedge\wedge}$, i.e. G/H is reflexive. Then $(G/H)^\wedge$ is reflexive, too, and H^o is reflexive because ϕ^H is a topological isomorphism.

The reflexivity of H follows from (14.6), (14.3) and (14.4). Consequently, $H^\wedge$ is reflexive, and $G^\wedge/H^o$ is reflexive becasue ϕ_H : $G^\wedge/H^o \to H^\wedge$ is a topological isomorphism due to (14.8). ■

(17.3) COROLLARY. Let G be a metrizable and complete nuclear group and let $(A_n)_{n=1}^\infty$ be a sequence of LCA groups. Then the product $G \times \prod_{n=1}^{\infty} A_n$ is a strongly reflexive group.

Proof. Naturally, $P := G \times \prod_{n=1}^{\infty} A_n$ is Čech-complete. We have to show that P is binuclear. That P is nuclear follows from (7.10) and

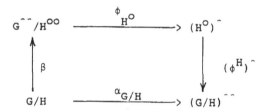

(7.6). The nuclearity of $P^{\wedge}$ is a consequence of (14.11), (16.1), (1.8) and (7.8):

$$P^{\wedge} = (G \times \prod_{n=1}^{\infty} A_n)^{\wedge} \sim G^{\wedge} \oplus \sum_{n=1}^{\infty} A_n^{\wedge}.$$

In view of (1.8) and (14.11), it suffices to prove that G is reflexive.

According to (9.7), we can find a metrizable and complete nuclear vector group F, a closed subgroup H of F and a closed subgroup Q of H, such that $G \sim H/Q$. From (15.7) and (16.5) it follows that F is a binuclear group. Hence, by (17.2), it is strongly reflexive. Finally, (17.1) (d) implies that H/Q is reflexive. ∎

We shall now give some examples of groups which are not strongly reflexive; they show that (17.3) cannot be much generalized. Denote by R^{ω} the countable product of real lines. We begin with the following technical result:

(17.4) LEMMA. Let M be a dense linear subspace of a metrizable locally convex space E. Then the identity mappings $E_{pc}^{\wedge} \to E_c^{\wedge} \to M_{pc}^{\wedge} \to M_c^{\wedge}$ are all topological isomorphisms.

Proof. It is enough to show that $id : M_c^{\wedge} \to E_{pc}^{\wedge}$ is continuous. Let X be an arbitrary precompact subset of E. We have to find a compact subset Y of M with $Y^{\circ} \subset X^{\circ}$. Naturally, there is a precompact subset A of M with $X \subset \bar{A}$. Next, there is a null sequence $(u_n)_{n=1}^{\infty}$ in M with $A \subset \overline{\text{conv}}\, \{u_n\}_{n=1}^{\infty}$ (see [80], p. 151). Obviously, the set

$$Y = \{tu_n : -1 \leq t \leq 1; \quad n = 1,2,\ldots\}$$

is compact. Let $\chi \in Y^{\circ}$. By (2.3), we have $\chi = \rho f$ for some $f \in E^*$. From (2.2) it follows that $f(u) = \chi(u)$ for $u \in Y$. Thus $|f(u)| \leq \frac{1}{4}$ for all $u \in Y$ and, consequently, for all $u \in \overline{\text{conv}}\, Y$. Since $X \subset \overline{\text{conv}}\, Y$, for each $u \in X$, we have $|\chi(u)| = |\rho f(u)| \leq |f(u)| \leq \frac{1}{4}$, which means that $\chi \in X^{\circ}$. ∎

(17.5) REMARK. Let E be a locally convex space and let $E_{\tau}^{\wedge}$ be the dual group (or, which is the same, the dual space) endowed with the topology of uniform convergence on compact convex sets. Then $id: E^{\wedge}_{\tau} \to E^{\wedge}_{c}$ need not be continuous. For instance, let E be the subspace of R^{ω} consisting of finite sequences. It is not hard to see that every compact convex subset of E is finite dimensional. Therefore $E^{\wedge}_{\tau}$ may identified with E, whereas $E^{\wedge}_{c}$ is simply the direct sum of real lines. Notice that the evaluation map is a topological isomorphism of E onto $(E^{\wedge}_{\tau})^{\wedge}_{\tau}$.

(17.6) REMARK. O.G. Smolyanov [84] proved that the space $\mathcal{D} = \mathcal{D}(R)$ of test functions on the real line contains a closed linear subspace L such that $\mathcal{D}/L$ is topologically isomorphic to a non-closed dense subspace M of R^{ω}. By (17.4), we have $M^{\wedge} \sim (R^{\omega})^{\wedge}$, and hence $M^{\wedge\wedge} \sim (R^{\omega})^{\wedge\wedge} \sim R^{\omega}$, so that M is not a reflexive group. Consequently, $\mathcal{D}$ is not strongly reflexive. Observe that $\mathcal{D}$ is a reflexive nuclear space, and the dual space $\mathcal{D}'$ is nuclear, too. Thus $\mathcal{D}$ is a reflexive binuclear group ((7.4) and (15.3)). This example shows that the assumption of the metrizability of G in (17.3) is essential.

Let $N = L^0$ be the annihilator of L in $\mathcal{D}'$. We may write

$$M \sim \mathcal{D}/L \sim \mathcal{D}''/L^{00} \sim (L^0)' = N' \sim N^{\wedge}.$$

Since M is not a reflexive group, it follows that α_N is not continuous (cf. (15.8)).

The space $\mathcal{D}(R)$ may be replaced here by the space $\mathcal{D}(\Omega)$ of test functions on an arbitrary open subset Ω of R^n because the latter one has a quotient topologically isomorphic to $\mathcal{D}(R)$ (see [90] or [85]). Other examples of spaces with quotients isomorphic to non-closed dense subspaces of R^{ω} are given in [91], Theorem 2 and [92], Theorem 7.

(17.7) EXAMPLE. Let ωR be the countable direct sum of real lines. We shall prove that the group $E = \omega R \times R^{\omega}$ is not strongly reflexive. Elements of E may be identified with left-hand finite sequences of real numbers. The family of sets of the form

$$\{(x_n) \in E : |x_n| < \varepsilon_n \text{ for } n \le n_o\},$$

where $n_o = 1,2,\ldots$ and $(\varepsilon_n)_{n \le n_o}$ is a sequence of positive numbers, is a base at zero in E. By (14.11), the dual group $E^{\wedge}$ may be identified with the space $R^{\omega} \times \omega R$ of right-hand finite sequences. Thus E is a reflexive binuclear group.

For each $n \in Z$, let $e_n \in E$ be the sequence with 1 in the n-th place and 0 elsewhere. Let e_n^*, $n \in Z$, be the continuous linear functional on E given by $e_n^*(e_m) = \delta_{mn}$. We shall identify the dual space E^* with $R^\omega \times \omega R$ in the usual way.

Let ϑ be some fixed irrational number. For each $m = 1,2,\dots$, set

$$u_m = e_{-m} + e_m, \qquad u_m^* = (1-\vartheta)^{-1}(e_{-m}^* - \vartheta e_m^*),$$

$$w_m = \vartheta e_{-m} + e_m, \qquad w_m^* = (1-\vartheta)^{-1}(e_m^* - e_{-m}^*).$$

Let K be the subgroup of E generated by the set $H = \{e_o\} \cup \{u_m\}_{m=1}^\infty \cup \{w_m\}_{m=1}^\infty$ and let K^* be the subgroup of E^* generated by the set $H^* = \{e_o^*\} \cup \{u_m^*\}_{m=1}^\infty \cup \{w_m^*\}_{m=1}^\infty$. A direct calculation shows that

$$K = \{u \in E : f(u) \in Z \text{ for all } f \in K^*\},$$

therefore K is a (weakly) closed subgroup of E. Observe that the linear mapping $\Phi : E \to E^*$ defined by the conditions

$$\Phi e_m = \begin{cases} (1-\vartheta)^{-1} e_{-m}^* & \text{when } m < 0 \\ e_o^* & \text{when } m = 0 \\ -(1-\vartheta)^{-1} e_{-m}^* & \text{when } m > 0 \end{cases}$$

is a topological isomorphism which carries K over onto K^*. Consequently, the mapping $\rho_{E^*}\Phi : E \to E^\wedge$ is a topological isomorphism which carries K over onto K^o. This means that $E^\wedge/K^o \sim E/K$.

Let $\dot{E}$ be the group E endowed with the topology induced by its canonical embedding into $R^\omega \times R^\omega$. It is not hard to see that $id : E/K \to \dot{E}/K$ is a topological isomorphism. Let $\psi : E \to E/K$ be the canonical projection. The set

$$H' = \{tu : t \in [0,1];\ u \in H\}$$

is obviously compact in $\dot{E}$. Consequently, $\psi(H')$ is a compact subset of E/K. We shall prove that

(1) $[\psi(H')]^{\circ}_{E/K} = \{0\}$.

Take any $\chi \in (E/K)^{\wedge}$ with $|\chi(\psi(H'))| \leq \frac{1}{4}$. Then $\chi\psi \in E^{\wedge}$, and, by (2.3), there is some $f \in E^*$ with $\rho f = \chi\psi$. From (2.2) we obtain

(2) $f(u) = \chi\psi(u)$ for all $u \in H'$.

Now, take any $v \in H$ and denote $I_v = \{tv : t \in [0,1]\}$. Then $\psi(I_v)$ is a subgroup of E/K; since $|\chi\psi(I_v)| \leq \frac{1}{4}$, it follows that $|\chi\psi(I_v)| = 0$ and (2) implies that $f(v) = 0$. Since $v \in H$ was arbitrary and H is linearly dense in E, we see that $f = 0$, which means that $\chi = 0$. This proves (1).

Let T^{H^*} be the group of all functions $f : H^* \to T$ endowed with the topology of pointwise convergence. Let $\phi : E/K \to T^{H^*}$ be the homomorphism given by the formula

$$\phi(\psi(u))(v^*) = \rho v^*(u) \qquad (u \in E, \quad v^* \in H^*)$$

(it can be shown that ϕ is a topological embedding). Take any $u = \sum_{n\in Z} x_n e_n \in E$ and set $f = \phi(\psi(u))$. Then, according to our definitions, for each $n \in Z$, we have

$$f(u_n^*) = \rho[(1 - \vartheta)^{-1}(x_{-n} - \vartheta x_n)],$$

$$f(w_n^*) = \rho[(1 - \vartheta)^{-1}(x_n - x_{-n})],$$

which can be written as

$$x_n = f(u_n^*) + f(w_n^*) + k_n + l_n,$$

$$x_{-n} = f(u_n^*) + \vartheta f(w_n^*) + k_n + \vartheta l_n,$$

for some $k_n, l_n \in Z$. Since $\Sigma x_n e_n \in E$, there is an index $n_o > 0$ such that, for each $n > n_o$, one has $x_{-n} = 0$, which is clearly possible if and only if

(3) $f(u_n^*) + \vartheta f(w_n^*) \in Z + \vartheta Z$.

The above argument shows that $\phi(E/K)$ consists of all those functions $f : H^* \to T$ which satisfy (3) for sufficiently large n. So, $\phi(E/K)$ is a non-closed dense subgroup of the compact group T^{H^*}. Now, (1) implies that $(E/K)^\wedge$ is discrete. Therefore, if E/K were reflexive, it would have to be compact and then $\phi(E/K)$ would be a compact subgroup of T^{H^*}. This proves that E/K cannot be reflexive. It is not hard to see that $(E/K)^{\wedge\wedge}$ is compact and $\alpha_{E/K}$ is a topological isomorphism of E/K onto a non-closed dense subgroup of $(E/K)^{\wedge\wedge}$ (cf. (14.7)).

From (14.8) we see that $\phi_K : E^\wedge/K^O \to K^\wedge$ is a topological isomorphism. Since $E^\wedge/K^O \sim E/K$, and E/K is not reflexive, it follows that K is not reflexive, either. It can be shown that α_K is an open, but not continuous, isomorphism of K onto $K^{\wedge\wedge}$, that $K^{\wedge\wedge}$ is discrete and that each precompact subset of K is finite. The homomorphism $\phi^K : (E/K)^\wedge \to K^O$ is not open; consequently, $\psi : E \to E/K$ is not compact-covering.

The above example shows that in (17.3) the countable product of LCA groups cannot be replaced by the countable direct sum.

(17.8) REMARK. By (17.3), the groups ωR and R^ω are both strongly reflexive. Therefore (17.7) shows that the product of two strongly reflexive groups need not be strongly reflexive. It is quite easy to show that the product of a strongly reflexive group and a compact or discrete one is strongly reflexive again; it is not known if the product of a strongly reflexive group and the real line must be strongly reflexive.

A standard argument shows that each closed linear subspace and each Hausdorff quotient space of $\omega(R^\omega)$ and $(\omega R)^\omega$ are reflexive (the meaning of the symbols is obvious). The author does not know if the same is true for the spaces $(\omega(R^\omega))^\omega$ and $\omega((\omega R)^\omega)$.

(17.9) REMARK. A subset of ωR is open if and only if it intersects each finite dimensional subspace in a relatively open set ([26], Proposition 1 or [93], Proposition 4 on p. 477). So, ωR is a k-space and R^ω is a k-space because it is metrizable. Nevertheless, $\omega R \times R^\omega$ is not a k-space; the function $(x_n)_{n\in Z} \to \sum_{n=1}^{\infty} x_n x_{-n}$ is k-continuous, but not continuous.

(17.10) REMARK. Let $G_n = R$ for all $n \in Z$, let $H_n = \{0\}$ for $n < 0$ and $H_n = R$ for $n \geq 0$. It is then clear that $\omega R \times R^{\omega} \sim \sum_{n \in Z} (G_n : H_n)$. So, (17.7) shows that countable reduced products of LCA groups need not be strongly reflexive. Nevertheless, if G is a reflexive binuclear group (in particular, a countable reduced product of LCA groups) and K is a closed subgroup of G, then $\phi_K : G^{\wedge}/K^{o} \to K^{\wedge}$ is a topological isomorphism, $\phi^K : (G/H)^{\wedge} \to K^{o}$ is an open isomorphism, α_K is an open isomorphism of K onto $K^{\wedge\wedge}$ and $\alpha_{G/K}$ is a topological embedding of G/K into $(G/K)^{\wedge\wedge}$. The proofs of these assertions are left to the reader.

(17.11) EXAMPLE. Let Ω be the first uncountable cardinal and, for each $\alpha < \Omega$, let G_α be the group $Z/2Z$. Let G be (algebraically) the group $\sum_{\alpha<\Omega} G_\alpha$. As a base at zero in G we take the family $\{H_\beta\}_{\beta<\Omega}$ where

$$H_\beta = \{(g_\alpha)_{\alpha<\Omega} : g_\alpha = 0 \text{ for all } \alpha < \beta\}.$$

Then G is a non-discrete Hausdorff group and every H_β is an open subgroup of G.

The groups G/H_β are discrete, and the canonical projections $\pi_\beta : G \to G/H_\beta$ induce a homomorphism π of G into the limit K of the inverse system

$$\{\pi_{\gamma\beta} : G/H_\gamma \to G/H_\beta;\ \ \beta \leq \gamma < \Omega\}$$

where $\pi_{\gamma\beta}$ are the canonical projections. Evidently, π is a topological embedding. As a matter of fact, G is complete and, therefore, π is a topological isomorphism of G onto K (see [72], Example 1.6 and [102], §5). So, G may be treated as a closed subgroup of the product $\prod_{\beta<\Omega} (G/H_\beta)$. We shall show that G is not reflexive; this means that the countable product of LCA groups in (17.3) cannot be replaced by an uncountable one (see, however, (17.14)).

The dual group $G^{\wedge}$ may be identified algebraically with the subgroup Q of the product $\prod_{\alpha<\Omega} G_\alpha^{\wedge}$, consisting of all elements $g^{\wedge} = (g_\alpha^{\wedge})_{\alpha<\Omega}$ with the property that there is some $\beta < \Omega$ (depending on $g^{\wedge}$)

such that $g_\alpha^\wedge = 0$ when $\beta \le \alpha$. It is not hard to see that every precompact subset of G is finite; therefore $G^\wedge$ may be identified with Q topologically. For each $\beta < \Omega$, let $g_\beta^\wedge = (g_{\beta\alpha}^\wedge)_{\alpha<\Omega}$ be the element of $\prod_{\alpha<\Omega} G_\alpha^\wedge$ with $g_{\beta\beta}^\wedge \neq 0$ and $g_{\beta\alpha}^\wedge = 0$ when $\alpha \neq \beta$. It is clear that the set $\{0\} \cup \{g_\beta^\wedge\}_{\beta<\Omega}$ is compact, and its quasi-convex hull in $G^\wedge$ is $G^\wedge$ itself. Consequently, $G^{\wedge\wedge}$ is discrete. Naturally, we may identify $G^{\wedge\wedge}$ with $\sum_{\alpha<\Omega} G_\alpha$. This means that α_G is an open, but not continuous, isomorphism of G onto $G^{\wedge\wedge}$.

The possibility of extending the Pontryagin-van Kampen duality theorem to non-locally compact abelian groups was investigated in several papers. A brief survey can be found in [38], in the notes after §§23-25. Here we shall restrict ourselves to a review of results connected with strong duality.

Let $(A_n)_{n=1}^\infty$ be a sequence of LCA groups. Set $G = \prod_{n=1}^\infty A_n$ and $H = \sum_{n=1}^\infty A_n$. Let P and Q be arbitrary closed subgroups of G and H, respectively.

Kaplan [49] proved that G and H are reflexive, and that there are canonical topological isomorphisms $G^\wedge \sim \sum_{n=1}^\infty A_n^\wedge$ and $H^\wedge = \prod_{n=1}^\infty A_n^\wedge$. Kaplan [50] proved that P is dually closed and dually embedded in G. Varopoulos [93] proved that Q is dually closed and dually embedded in H. Kaplan [50] proved that limits of direct and inverse sequences of LCA groups are reflexive. Similar results were independently obtained by Vilenkin [99].

Varopoulos [93] showed that $\phi^Q : (H/Q)^\wedge \to Q^o$ and $\phi_H : H^\wedge/Q^o \to Q^\wedge$ are topological isomorphisms. Noble [72] proved that P is reflexive. He introduced a special class of topological groups, the so-called k--groups (see (14.5)) and proved that if K is a k-group, then α_K is continuous. It can be shown that $P, Q, G/P$ and H/Q are all k-groups. Brown et al. [26] showed that G is strongly reflexive if A_1 is compact and A_n is isomorphic to R or to Z for $n \ge 2$ (or, dually, if A_1 is discrete and A_n is isomorphic to R or to T for $n \ge 2$). Venkataraman [95] proved that P is reflexive, and that $\phi_P : G^\wedge/P^o \to P^\wedge$ is a topological isomorphism.

(17.12) REMARK. As a special case of (17.3) we obtain the following proposition:

(*) countable products of LCA groups are strongly reflexive.

It turns out that (*) is an almost immediate consequence of the results of [50] and [93] (see [12]). It was pointed out to the author by E. Martín-Peinador that, as a matter of fact, (*) is a direct consequence of the results of [50] only, which allows one to obtain a relatively elementary proof of (*). Another elementary proof of (*) can be obtained by following the way of [26].

(17.13) REMARK. Let A be an open subgroup of a Hausdorff abelian group G. Then A is dually closed and dually embedded in G ([72], Lemma 3.3). Venkataraman [94],Corollary 6.3, proved that if G is reflexive, then A is reflexive, too (for k-groups, this was proved by Noble [72], Corollary 3.4). It can be shown that G is reflexive (resp. strongly reflexive) if and only if A is reflexive (resp. strongly reflexive); the proof will be given in [29].

A group locally isomorphic to a reflexive group need not be reflexive itself (see (15.3) and (5.1) or (5.3)). The author does not know whether a group locally isomorphic to a strongly reflexive one must be strongly reflexive.

(17.14) REMARK. Negrepontis [71] proved that direct and inverse limits of compactly generated LCA groups are reflexive (cf. (17.11)). There is, however, an error in the proof of Proposition 4.4 on p. 250. The author does not know whether closed subgroups of uncountable direct sums of real lines are dually closed or dually embedded. Also, it remains an open question whether closed subgroups or Hausdorff quotients of uncountable products of R´s or of Z´s are reflexive. Brown et al. [26] proved that each closed subgroup of the countable product of R´s can be written in the form

$$\{(x_n) : x_n = 0 \text{ for } n \in A \text{ and } x_n \in Z \text{ for } n \in B\}$$

for some disjoint subsets A,B of $\{1,2,\dots\}$. Most likely, this result cannot be extended to uncountable products.

(17.15) NOTE. The material of this section is new, with the following exceptions. Strong reflexivity was introduced in [26]. Parts (a) and (b) of (17.1) were communicated to the author by M.J. Chasco and

E. Martín-Peinador. Parts (c) and (d) of (17.1) are, in fact, taken from [26]. The special case of (17.3) when G is a nuclear Fréchet space and all groups A_n are trivial appears in [8]. Example (17.11) is taken from [72]; as a matter of fact, this is the example given by Leptin in [61].

18. Groups with boundedness

One of the main obstacles in extending Pontryagin duality to non-locally compact groups is that the quotient homomorphism $G \to G/H$ need not be compact-covering (cf. (17.6) and (17.7)). To avoid this difficulty, N.Ya. Vilenkin [98] introduced the so-called groups with boundedness. In this section we give an outline of Vilenkin's theory and show how groups with boundedness allow one to obtain an analogue of (17.2) for binuclear groups which are not Čech-complete. The main results are (18.4) and (18.7).

Let G be an abelian group. By a <u>boundedness</u> on G we mean a family of subsets of G, called <u>bounded sets</u>, satisfying the following conditions:

1) if X is bounded, so is $-X$;

2) subsets of bounded sets are bounded;

3) if X,Y are bounded, so are $X \cup Y$ and $X + Y$;

4) finite sets are bounded.

For instance, the family of all finite (resp. precompact or relatively compact) subsets of a topological group forms a boundedness. Another example is the family of all subsets of a topological vector space which are bounded in the usual sense i.e. absorbed by neighbourhoods of zero. Equipped with this boundedness, a topological vector space E will be denoted by E_σ.

Let G be a group with boundedness and H its subgroup. The boundedness on G induces the boundedness on H and on G/H. As bounded subsets of H we take the sets $H \cap X$ where X is a bounded subset of G. Bounded subsets of G/H are defined as the canonical images of bounded subsets of G.

If L is a closed linear subspace of a topological vector space E, then it may happen that $(E/L)_\sigma \neq E_\sigma/L$; for instance, in the notation of (17.6), there are bounded subsets of $\mathcal{D}/L$ which are not canonical images of bounded subsets of $\mathcal{D}$.

Let G be a topological group with boundedness. We say that G is a Q-group if it is locally quasi-convex and if quasi-convex hulls of bounded sets are bounded. If E is a locally convex space, then every closed, convex and symmetric subset of E is quasi-convex (see the proof of (2.4)); therefore E_σ is a Q-group. It is obvious that any subgroup of a Q-group is a Q-group. However, a Hausdorff quotient of a Q-group need not be locally quasi-convex (see (5.3)).

Let $\phi : G \to H$ be a homomorphism of groups with boundedness. We say that ϕ is bounded if the images of bounded sets are bounded. We say that ϕ is bounding if to each bounded $X \subset H$ there corresponds some bounded $Y \subset G$ with $X \cap \phi(G) \subset \phi(Y)$. Finally, ϕ is called bi-bounded if it is both bounded and bounding. Bi-bounded topological isomorphisms will be called b-isomorphisms.

Let G be a topological group with boundedness. The dual group, denoted by G^d, is defined in the following way. As a set, G^d consists of all continuous characters of G. As a base at zero in G^d we take the polars of bounded subsets of G. The boundedness on G^d consists of equicontinuous sets. It follows immediately from our definitions that G^d is a Q-group. Notice that if $\phi : G \to H$ is a b-isomorphism, then the dual mapping $\phi^d : H^d \to G^d$, defined in the usual way, is a b-isomorphism, too.

The evaluation map $G \to G^{dd}$ will be denoted by β_G. We say that G is an involutive group if β_G is a b-isomorphism of G onto G^{dd}.

(18.1) PROPOSITION. Let G be a separated Q-group. Then β_G is a b-isomorphism of G onto $\beta_G(G)$.

This is a direct consequence of our definitions.

(18.2) PROPOSITION. A locally convex space E is semi-reflexive if and only if E_σ is an involutive group.

Proof. By (2.3), we have the topological isomorphism $\rho_E : E_b^* \to (E_\sigma)^d$ which, again by (2.3), allows us to identify algebraically $(E_b^*)^*$ with $(E_\sigma)^{dd}$. Thus, E is semi-reflexive if and only if im $\beta_{E_\sigma} = (E_\sigma)^{dd}$. Now, it suffices to apply (18.1) because E_σ is a Q-group. ∎

(18.3) PROPOSITION. A dually closed, dually embedded subgroup of an involutive group is involutive.

Proof. Let H be a dually closed, dually embedded subgroup of an involutive group G. In view of (18.1), it is enough to show that $\operatorname{im} \beta_H = H^{dd}$. Take an arbitrary continuous character ξ of H^d. The canonical homomorphism $\phi : G^d \to H^d$ is continuous, therefore $\xi\phi$ is a continuous character of G^d. It is obvious that $\xi\phi \in H^{oo}$. Since H is dually closed in G, we have $\beta_G(H) = H^{oo}$ (cf. (14.2)). So, there is some $h \in H$ with $\beta_G(h) = \xi\phi$ and it is not hard to verify that $\beta_H(h) = \xi$ (cf. the proof of (14.6)). ∎

Let $G = \sum_{i\in I} (G_i : H_i)$ be a reduced product of groups with boundedness. We introduce boundedness into G in the following way. For each $i \in I$, let $\pi_i : G \to G_i$ be the canonical projection. A subset X of G is bounded if $\pi_i(X)$ is a bounded subset of G_i for all $i \in I$ and if $\pi_i(X) \subset H_i$ for all but finitely many i.

(18.4) PROPOSITION. Let $G = \sum_{i\in I} (G_i : H_i)$ be a reduced product of topological groups with boundedness. Suppose that H_i is dually closed in G_i for almost all i. Then G^d is canonically b-isomorphic to $\sum_{i\in I} (G_i^d : H_i^o)$. Consequently, if all groups G_i are involutive, then G is involutive, too.

Proof. Take any $\chi \in G^d$. Since χ is continuous, one has $|\chi(U)| \leq \frac{1}{4}$ for some $U \in N_o(G)$. According to the definition of the topology on G, we may assume that

(1) $\qquad U = \{(g_i) \in G : g_i \in U_i \text{ for all } i \in I\}$

where $U_i \in N_o(G_i)$ for all i and $H_i \subset U_i$ for almost all i. Let $\chi_i = \chi_{|G_i} \in G_i^d$ for each $i \in I$. Then $\chi_i \in H_i^o$ for almost all i, which means that

$$(\chi_i)_{i\in I} \in G' := \sum_{i\in I} (G_i^d : H_i^o).$$

Consider the mapping $\phi : G^d \to G'$ given by $\phi(\chi) = (\chi_i)_{i\in I}$. It is clear that ϕ is an algebraical isomorphism of G^d onto G'. We shall prove that ϕ is a b-isomorphism.

1° ϕ <u>is continuous</u>. Take any $W \in N_o(G')$. According to the definition of the topology on G', we may asume that

$$W = \{(\chi_i) \in G' : \chi_i \in W_i \text{ for all } i \in I\}$$

where $W_i \in N_o(G_i^d)$ for all i and $H_i^o \subset W_i$ for almost all i. For each $i \in I$, there exists a bounded subset X_i of G_i with $X_i^o \subset W_i$. If $X_i \not\subset H_i$, then $H_i^o \not\subset X_i^o$ because H_i is dually closed in G_i. This means that $X_i \subset H_i$ for almost all i. Then the set

(2) $$X = \{(g_i) \in G : g_i \in X_i \text{ for all } i \in I\}$$

is bounded in G. We have $X^o \in N_o(G^d)$ and it is clear that $\phi(X^o) \subset W$.

2° ϕ <u>is open</u>. Take any $V \in N_o(G^d)$. There is a bounded subset X of G with $X^o \subset V$. We may assume that X has form (2) where X_i is a bounded subset of G_i for all i and $X_i \subset H_i$ for almost all i. We may write

$$X^o = \{(\chi_i) \in G^d : \chi_i \in X_i^o \text{ for all } i \in I\}.$$

If $X_i \subset H_i$, then $H_i^o \subset X_i^o$. So, we have $H_i^o \subset X_i^o$ for almost all i, which means that $\phi(X^o) \in N_o(G')$.

3° ϕ <u>is bounded</u>. Let X be a bounded subset of G^d. There is some $U \in N_o(G)$ with $X \subset U^o$. We may assume that U has form (1). Then

$$U^o = \{(\chi_i) \in G^d : \chi_i \in U^o \text{ for all } i \in I\}.$$

If $H_i \subset U_i$, then $U_i^o \subset H_i^o$; since this holds for almost all i, it follows that $\phi(U^o)$ is a bounded subset of G'.

4° ϕ <u>is bounding</u>. Let Y be a bounded subset of G'. We may assume that

$$Y = \{(\chi_i) \in G' : \chi_i \in Y_i \text{ for all } i \in I\}$$

where Y_i is a bounded subset of G_i^d for all i and $Y_i \subset H_i^o$ for all i outside a certain finite subset J of I. To each $i \in I$

there corresponds some $U_i \in N_o(G_i)$ with $Y_i \subset U_i^o$. Let A be the subset of G consisting of all sequences (g_i) such that $g_i \in U_i$ for $i \in J$ and $g_i \in U_i + H_i$ for $i \notin J$. Then $A \in N_o(G)$ and A^o is the bounded subset of G^d consisting of all sequences (χ_i) such that $\chi_i \in U_i^o$ for $i \in J$ and $\chi_i \in (U_i + H_i)^o$ for $i \notin J$. Since H_i is a subgroup of G_i, we have $(U_i + H_i)^o = U_i^o \cap H_i^o$. This implies that $Y_i \subset (U_i + H_i)^o$ for $i \notin J$. Hence $Y \subset \phi(A^o)$. ■

An involutive group G is called <u>strongly involutive</u> if every closed subgroup and every Hausdorff quotient group of G and of G^d is involutive.

(18.5) LEMMA. Let G be a topological group with boundedness. For each closed subgroup H of G, the canonical mapping $\phi^H : (G/H)^d \to H^o$ is a b-isomorphism.

The verification of this simple fact is left to the reader.

(18.6) PROPOSITION. Let H be a closed subgroup of a strongly involutive group G. Then

(a) H is dually closed and dually embedded in G;

(b) the canonical mappings $\phi_H : G^d/H^o \to H^d$ and $\phi^H : (G/H)^d \to H^o$ are b-isomorphisms;

(c) the groups H and G/H are strongly involutive.

The proof is similar to that of (17.1). We leave it to the reader.

Let G be an abelian topological group. By $[G]$ we shall denote the group G equipped with the boundedness consisting of precompact sets.

(18.7) THEOREM. Let G be a reflexive group. Suppose that both the groups G and $G^\wedge$ are nuclear and complete. Then the group $[G]$ is strongly involutive.

The assumptions of (18.7) are satisfied, in particular, for $G = \mathcal{D}(\Omega)$ or $G = \omega R \times R^\omega$ (cf. (17.6) and (17.7)).

Proof. It follows directly from our assumptions that the identity mappings $i : [G]^d \to [G^\wedge]$ and $j : [G^\wedge]^d \to [G^{\wedge\wedge}]$ are b-isomorphisms. Then the sequence

$$[G] \xrightarrow{\alpha_G} [G^{\wedge\wedge}] \xrightarrow{j^{-1}} [G^\wedge]^d \xrightarrow{i^d} [G]^{dd}$$

shows that $[G]$ is involutive. By (8.6) and (8.3), closed subgroups of $[G]$ and of $[G]^d$ are dually closed and dually embedded. Hence, by (18.3), each closed subgroup of $[G]$ and of $[G]^d$ is involutive. It remains to show that Hausdorff quotient groups of $[G]$ and of $[G]^d$ are involutive.

So, let H be a closed subgroup of $[G]$. The canonical mapping $\gamma : [G]/H \to [G]^{dd}/H^{oo}$ is a b-isomorphism (see (14.2)). Next, (14.8) implies that the canonical mapping $\phi_{H^o} : [G]^{dd}/H^{oo} \to (H^o)^d$ is a topological isomorphism. It is clear that ϕ_{H^o} is bounded; that it is follows from (8.2). Finally, (18.5) says that $\phi^H : ([G]/H)^d \to H^o$ is a b-isomorphism. Now, the sequence

$$[G]/H \xrightarrow{\gamma} [G]^{dd}/H^{oo} \xrightarrow{\phi_{H^o}} (H^o)^d \xrightarrow{(\phi^H)^d} ([G]/H)^{dd}$$

shows that $[G]/H$ is involutive. The proof that Hausdorff quotients of $[G]^d$ are involutive is similar. ▪

(18.8) REMARK. If E is a non-reflexive Banach space, then E is a reflexive group, but E_σ is not involutive (see (15.2) and (18.2)). Therefore it is interesting that a metrizable locally convex space E is a strongly reflexive group if and only if E_σ is strongly involutive, namely, if and only if E is a nuclear Fréchet space. This follows from (17.3), (18.7) and (6.1).

(18.9) NOTE. The material of this section is taken from [98]. Theorem (18.7) is new.

BIBLIOGRAPHY

1. R. Arens, Duality in linear spaces, Duke Math. J. **14** (1947), 787-793.
2. A. Arkhangelskiĭ, Open and close-to-open mappings. Relations among spaces, Trudy Moskov. Mat. Obsch. **15** (1966), 181-223.
3. K. Ball, Some remarks on the geometry of convex sets, GAFA Seminar ´86-87, Springer Lecture Notes in Math. **1317** (1988), 224-- 231.
4. M. Banaszczyk and W. Banaszczyk, Characterization of nuclear spaces by means of additive subgroups, Math. Z. **186** (1984), 125--133.
5. W. Banaszczyk, On the existence of unitary representations of commutative nuclear Lie groups, Studia Math. **76** (1983), 175-181.
6. ——, On the existence of exotic Banach-Lie groups, Ann. Math. **264** (1983), 485-493.
7. ——, Closed subgroups of nuclear spaces are weakly closed, Studia Math. **80** (1984), 119-128.
8. ——, Pontryagin duality for subgroups and quotients of nuclear spaces, Math. Ann. **273** (1986), 653-664.
9. ——, On the existence of commutative Banach-Lie groups which do not admit continuous unitary representation, Coll. Math. **52** (1987), 113-118.
10. ——, The Steinitz theorem on rearrangements of series for nuclear spaces, J. Reine Angew. Math. **403** (1990), 187-200.
11. ——, Polar lattices from the point of view of nuclear spaces, Rev. Mat. Univ. Complut. Madrid 2 (special issue) (1989), 35-46.
12. ——, Countable products of LCA groups: their closed subgroups, quotients and duality properties, Colloq. Math. **59** (1990), 53-57.
13. ——, A Beck-Fiala-type theorem for euclidean norms, to appear in European J. Combin. **11** (1990).
14. ——, Rearrangements of series in non-nuclear spacess, in preparation.
15. W. Banaszczyk and J. Grabowski, Connected subgroups of nuclear spaces, Studia Math. **78** (1984), 161-163.

16. I. Bárány, Rearrangements of series in infinite dimensional spaces, Mat. Zametki **46** (1989), no. 6, 10-17 (Russian), translated as Math. Notes (preprint 1983).

17. I. Bárány and V.S. Grinberg, On some combinatorial questions in finite dimensional spaces, Linear Algebra Appl. **41** (1981), 1-9.

18. A.O. Barut and R. Rączka, Theory of group representations and applications, PWN - Polish Scientific Publishers, Warszawa 1977.

19. J. Beck, Balancing families of integer sequences, Combinatorica **1** (1981), 209-216.

20. J. Beck and T. Fiala, Integer-making theorems, Discrete Appl. Math. **3** (1981), 1-8.

21. J. Beck and J. Spencer, Integral approximation sequences, Math. Programming **30** (1984), 88-98.

22. W. Blaschke, Über affine Geometry XI: lösing der "Vierpunktproblems" von Sylvester aus der Theorie der geometrischen Wahrsdeinlichkeiten. Leipziger Berichte **69** (1917), 436-453.

23. N. Bourbaki, Éléments de mathematique. Première partie. Les structures fondamentales de l´analyse. Livre III. Topologie générale. Chap. III. Groupes topologiques (Théorie élémentaire). Chap. 4. Nombres réels. 3 ed., Hermann, Paris 1960.

24. J. Bourgain and V.D. Milman, New volume ratio properties for convex symmetric bodies in R^n, Invent. Math. **88** (1987), 319-340.

25. J. Braconnier, Sur les groupes topologiques localement compacts, J. Math. Pures Appl. **27** (1948), 1-85.

26. R. Brown, P.J. Higgins and S.A. Morris, Countable products and sums of lines and circles: their closed subgroups, quotients and duality properties, Math. Proc. Cambridge Philos. Soc. **78** (1975), 19-32.

27. H. - P. Butzmann, Pontrjagin-Dualität für topologische Vektorräume, Arch. Math. (Basel) **28** (1977), 632-637.

28. J.W.S. Cassels, An introduction to the geometry of numbers, Springer-Verlag, Berlin 1959.

29. M.J. Chasco and E. Martín-Peinador, Open subgroups and Pontryagin duality, to appear in Math. Z.

30. I.M. Gelfand and N.Ya. Vilenkin, Generalized functions, vol. 4, Applications of harmonic analysis, Academic Press, New York - - London 1964.

31. M. Gromov and V.D. Milman, Brunn theorem and a concentration of volume of convex bodies, GAFA Seminar Notes, Tel Aviv University, Israel 1983-1984, Exp. V, 12 pp.

32. V.S. Grinberg and S.V. Sevastyanov, Value of the Steinitz constant, Funktsional. Anal. i Prilozhen. **14** (1980), no. 2, 56-57 (Russian), translated as Functional Anal. Appl. **14** (1980), 125-126.

33. P.M. Gruber and C.G. Lekkerkerker, Geometry of numbers, North - - Holland, Amsterdam 1987.

34. I. Halperin, Sums of series, permitting rearrangements, C.R. Math. Rep. Acad. Sci. Canada **8** (1986), 87-102.

35. G.H. Hardy, J.E. Littlewood and G. Pólya, Inequalities, Cambridge University Press, Cambridge 1952.

36. J. Hastad, Dual vectors and lower bounds for the nearest lattice point problem, Combinatorica **8** (1988), 75-81.

37. W. Herer and J.P.R. Christensen, On the existence of pathological submeasures and the construction of exotic topological groups, Math. Ann. **213** (1975), 203-210.

38. E. Hewitt and K.A. Ross, Abstract harmonic analysis, vol. I, Springer-Verlag, Berlin 1963.

39. E. Hille and R.S. Phillips, Functional analysis and semi-groups, Amer. Math. Soc., Providence 1957.

40. C.J. Himmelberg, Measure relations, Fund. Math. **87** (1975), 53-72.

41. R.C. Hooper, Topological groups and integer-valued norms, J. Funct. Anal. **2** (1968), 243-257.

42. F. John, Polar correspondence with respect to convex region, Duke Math. J. 3, No. **2**, University of Kentucky (1937), 355-369.

43. V.M. Kadets and M.I. Kadets, Rearrangements of series in Banach spaces, Tartu Gos. Univ., Tartu 1988 (Russian).

44. M.I. Kadets and K. Woźniakowski, On series whose permutations have only two sums, Bull. Polish Acad. Sci. Math. **37** (1989), 15-21.

45. V.M. Kadets, On a problem of S. Banach (problem 106 from "The Scottish Book"), Funktsional. Anal. i Prilozhen. **20** (1986), 74-75 (Russian), translated as Functional Anal. Appl. **20** (1986), 317--319.

46. ——, Series permutation in infinite dimensional spaces (main results and open problems), C.R. Math. Rep. Acad. Sci. Canada **11** (1989), 151-164.

47. R.V. Kadison and J.R. Ringrose, Fundamentals of the theory of operator algebras, vol. I Elementary theory, vol. II Advanced theory, Academic Press, New York 1983, 1986.

48. N. Kalton, Subseries convergence in topological groups and vector spaces, Israel J. Math. **10** (1971), 402-412.

49. S. Kaplan, Extensions of the Pontrjagin duality I: infinite products, Duke Math. J. **15** (1948), 649-658.

50. ——, Extensions of the Pontrjagin duality II: direct and inverse limits, ibid. **17** (1950), 419-435.

51. Y. Katznelson and O.C. McGehee, Conditionally convergent series in R^∞, Michigan Math. J. **21** (1974), 97-106.

52. J.L. Kelley, General topology, Van Nostrand, New York 1955.

53. A.A. Kirillov, Elements of the theory of representations, Springer-Verlag, Berlin 1976.

54. J. Kisyński, On the generation of tight measures, Studia Math. **30** (1968), 141-151.

55. Y. Komura, Some examples on linear topological spaces, Math. Ann.

153 (1984), 150-162.

56. S. Kwapień, On the form of a linear operator in the space of all measurable functions, Bull. Acad. Polon. Sci. Sér. Sci. Math. Astronom. Phys. **21** (1973), 951-954.

57. S.H. Kye, Pontryagin duality in real linear topological spaces, Chinese J. Math. **12** (1984), no. 2, 129-136.

58. J.C. Lagarias, H.W. Lenstra, Jr. and C.P. Schnorr, Korkin-Zolotarev bases and successive minima of a lattice and its reciprocal lattice, preprint 1989.

59. D.G. Larman and C.A. Rogers, The existence of a centrally symmetric convex body with central sections that are unexpectedly small, Mathematika **22** (1975), 164-175.

60. H. Leptin, Bemerkung zu einem Satz von S. Kaplan, Arch. Math. **6** (1955), 139-144.

61. ——, Zur Dualitätstheorie projektiver Limites abelscher Gruppen, Abh. Math. Sem. Univ. Hamburg **19** (1955), 264-268.

62. A. Mądrecki, Minlos theorem in p-adic locally convex spaces, preprint.

63. K. Maurin, General eigenfunction expansions and unitary representations of topological groups, PWN - Polish Scientific Publishers, Warszawa 1968.

64. S. Mazur, On the convergence of lacunary polynomials, Studia Math. **89** (1988), 75-78.

65. V.D. Milman and A. Pajor, Isotropic position and inertia ellipsoids and zonoids of the unit ball of a normed n-dimensional space, preprint IHES, May 1989.

66. J. Milnor and D. Husemoller, Symmetric bilinear froms, Springer-Verlag, Berlin 1973.

67. R.A. Minlos, Generalized stochastic processes and their extension to the measure, Trudy Moskov. Mat. Obshch. **8** (1959), 497-518 (Russian).

68. A.F. Monna, Analyse non-archimedienne, Springer-Verlag, Berlin 1970.

69. S.A. Morris, Pontryagin duality and the structure of locally compact abelian groups, Cambridge University Press, Cambridge 1977.

70. D.H. Muschtari, Some general questions of the theory of probability measures in linear spaces, Teor. Veroyatnost. i Primenen. **18** (1973), 66-77 (Russian).

71. J.W. Negrepontis, Duality in analysis from the point of view of triples, J. Algebra **19** (1971), 228-253.

72. N. Noble, k-groups and duality, Trans. Amer. Math. Soc. **151** (1970), 551-561.

73. W. Orlicz, Divergenz von allgemeinen Orthogonalreihen, Studia Math. **4** (1933), 27-32.

74. D.V. Pecherskiĭ, Series permutations in Banach spaces and rearrangements of signs, Mat. Sb. (N.S.) **135** (177) (1988), 24 - 35

(Russian), translated as Math. USSR-Sb.

75. A. Pietsch, Nuclear locally convex spaces, Springer-Verlag, Berlin 1972.

76. ——, Operator ideals, VEB Deutscher Verlag der Wissenschaften, Berlin 1978.

77. G. Pólya and G. Szegö, Problems and theorems in analysis, vol. II, Springer-Verlag, Berlin 1976.

78. D.A. Raikov, On B-complete topological vector spaces, Studia Math. **31** (1968), 295-306 (Russian).

79. S. Rolewicz, Metric linear spaces, second enlarged edition, PWN-Polish Scientific Publishers, Warszawa 1984.

80. H.H. Schaefer, Topological vector spaces, Springer-Verlag, Berlin 1971.

81. The Scottish Book, ed. by R. Daniel Mauldin, Birkhauser, Boston 1981.

82. S.J. Sidney, Weakly dense subgroups of Banach spaces, Indiana Univ. Math. J. **26** (1977), 981-986.

83. M.F. Smith, The Pontrjagin duality theorem in linear spaces, Ann. of Math. **56** (1952), 248-253.

84. O.G. Smolyanov, The spaces D is not hereditarily complete, Izv. Akad. Nauk SSSR Ser. Mat. **35** (1971), 628-696 (Russian), translated as Math. USSR-Izv. **5** (1971), 696-710.

85. ——, Analysis on topological vector spaces and its applications, Moscow 1979 (Russian).

86. J. Spencer, Six standard deviations suffice, Trans. Amer. Math. Soc. **289** (1985), 679-705.

87. E. Steinitz, Bedingt konvergente Reihen und konvexe Systeme, J. Reine Angew. Math. **143** (1913), 128-175.

88. S. Troyanski, On conditionally convergent series in certain F--spaces, Teor. Funtsiĭ, Funktsional. Anal. i Prilozhen. **5** (1967), 102-107 (Russian).

89. S.M. Ulam, A collection of mathematical problems, Interscience Publishers, New York 1960.

90. M. Valdivia, The spaces of distributions $\mathcal{D}'(\Omega)$ is not B_r-complete, Math. Ann. **211** (1974), 145-149.

91. ——, The space $\mathcal{D}(\Omega)$ is not B_r-complete, Ann. Inst. Fourier (Grenoble) **27** (1977), 29-43.

92. ——, (F)-spaces and strict (LF)-spaces, Math. Z. **195** (1987), 345--364.

93. N.Th. Varopoulos, Studies in harmonic analysis, Math. Proc. Cambridge Philos. Soc. **60** (1964), 465-516.

94. R. Venkataraman, Extensions of Pontryagin duality, Math. Z. **143** (1975), 105-112.

95. ——, A characterization of Pontryagin duality, Math. Z. **149** (1976), 109-119.

96. N.Ya. Vilenkin, Direct decompositions of topological groups I, Mat. Sb. **19** (61) (1946), 85-154 (Russian).

97. ——, Theory of topological groups. II. Direct products. Direct sums of groups of rank 1. Locally bicompact abelian groups. Fibered and weakly separable groups. Uspekhi Mat. Nauk **5** no.4 (38) (1950), 19-74 (Russian).

98. ——, Theory of characters of topological groups with a boundedness given, Izv. Akad. Nauk SSSR Ser. Mat. **15** (1951), 439-462 (Russian).

99. ——, Direct and inverse spectra of topological groups and their character theory, ibid., 503-532 (Russian).

100. A. Wald, Bedingt konvergente Reihen von Vektoren im R_ω, Ergebn. math. Kolloqu. **5** (1933), 13-14.

101. ——, Reihen in topologischen Gruppen, Ergebn. math. Kolloqu. **5** (1933), 14-16.

102. ——, L'intégration dans les groupes topologiques et ses applications, Actualités Sci. et Indust. **869, 1145**, Hermann & Cie., Paris 1940, 1951.

103. ——, Basic number theory, Springer-Verlag, Berlin 1967.

104. Y. Yang, On a generalization of Minlos theorem, Fudan Xuebao **20** (1981), no. 1, 31-37 (Chinese).

105. A. Zygmund, Trigonometric series, 2nd ed., vol. I, Cambridge University Press, Cambridge 1959.

INDEX OF SYMBOLS

(u,v)

$d(u,A)$

B_E, $B(E)$

$d_k(\Phi : E \to F)$, $d_k(\Phi)$

B_p, $B(p)$

E_p

ψ_p

Λ_{pq}

Section 3: B_n

vol_n

ω_n

$d(L)$

L^*

$\mu(L,U)$

$\lambda_k(L,U)$

D

$\xi_1 \leq \ldots \leq \xi_n$

Section 4: $L^2_C(X,\mu)$, $L^2_C(0,1)$

$L^\infty_C(X,\mu)$, $L^\infty_C(0,1)$

$L^0_C(X,\mu)$, $L^0_C(0,1)$

$L^p_A(X,\mu)$, $L^p_A(0,1)$

$e^{2\pi i\theta}$

Section 5: c_o

l^p

e_n

$\{f\}$

L^0

Section 6: $\frac{|U|}{|W|}$

$v_k(\Phi : E \to F)$, $v_k(\Phi)$

$h_k(\Phi : E \to F)$, $h_k(\Phi)$

Section 7: $c_o(N,G)$

Section 8: R^ω

$\mathcal{G}(E,F)$

Section 9: $w_o(G)$

Section 10: $\mathcal{S}(\Sigma g_i;G)$, $\mathcal{S}(\Sigma g_i)$

$\mathcal{C}(\Sigma g_i;G)$, $\mathcal{C}(\Sigma g_i)$

$\mathcal{A}(\Sigma g_i;G)$, $\mathcal{A}(\Sigma g_i)$

$\Gamma(\Sigma u_i)$

$\Gamma_0(\Sigma u_i)$

$\beta(\Phi)$, $\phi(\Phi)$, $\sigma(\Phi)$

Section 11: $\mathcal{B}(X)$

μ_π

$\hat{\mu}$

Section 14: α_G

ϕ_H, ϕ^H

Section 17: ωR, R^ω

Section 18: G^d

E_σ

$[G]$

ϕ^d

β_G

SUBJECT INDEX

Lecture Notes in Mathematics

For information about Vols. 1–1272
please contact your bookseller or Springer-Verlag

Vol. 1273: G.-M. Greuel, G. Trautmann (Eds.), Singularities, Representation of Algebras, and Vector Bundles. Proceedings, 1985. XIV, 383 pages. 1987.

Vol. 1274: N.C. Phillips, Equivariant K-Theory and Freeness of Group Actions on C*-Algebras. VIII, 371 pages. 1987.

Vol. 1275: C.A. Berenstein (Ed.), Complex Analysis I. Proceedings, 1985–86. XV, 331 pages. 1987.

Vol. 1276: C.A. Berenstein (Ed.), Complex Analysis II. Proceedings, 1985–86. IX, 320 pages. 1987.

Vol. 1277: C.A. Berenstein (Ed.), Complex Analysis III. Proceedings, 1985–86. X, 350 pages 1987.

Vol. 1278: S.S. Koh (Ed.), Invariant Theory. Proceedings, 1985. V, 102 pages. 1987.

Vol. 1279: D. Iesan, Saint Venant's Problem. VIII, 162 Seiten. 1987.

Vol. 1280: E. Neher, Jordan Triple Systems by the Grid Approach. XII, 193 pages. 1987.

Vol. 1281: O.H. Kegel, F. Menegazzo, G. Zacher (Eds.), Group Theory. Proceedings, 1986. VII, 179 pages. 1987.

Vol. 1282: D.E. Handelman, Positive Polynomials, Convex Integral Polytopes, and a Random Walk Problem. XI, 136 pages. 1987.

Vol. 1283: S. Mardesic', J. Segal (Eds.), Geometric Topology and Shape Theory. Proceedings, 1986. V, 261 pages. 1987.

Vol. 1284: B.H. Matzat, Konstruktive Galoistheorie. X, 286 pages. 1987.

Vol. 1285: I.W. Knowles, Y. Saito (Eds.), Differential Equations and Mathematical Physics. Proceedings, 1986. XVI, 499 pages. 1987.

Vol. 1286: H.R. Miller, D.C. Ravenel (Eds.), Algebraic Topology. Proceedings, 1986. VII, 341 pages. 1987.

Vol. 1287: E.B. Saff (Ed.), Approximation Theory, Tampa, Proceedings, 1985–1986. V, 228 pages. 1987.

Vol. 1288: Yu.L. Rodin, Generalized Analytic Functions on Riemann Surfaces. V, 128 pages, 1987.

Vol. 1289: Yu.I. Manin (Ed.), K-Theory, Arithmetic and Geometry. Seminar, 1984–1986. V, 399 pages. 1987.

Vol. 1290: G. Wüstholz (Ed.), Diophantine Approximation and Transcendence Theory. Seminar, 1985. V, 243 pages. 1987.

Vol. 1291: C. Mœglin, M.-F. Vignéras, J.-L. Waldspurger, Correspondances de Howe sur un Corps p-adique. VII, 163 pages. 1987.

Vol. 1292: J.T. Baldwin (Ed.), Classification Theory. Proceedings, 1985. VI, 500 pages. 1987.

Vol. 1293: W. Ebeling, The Monodromy Groups of Isolated Singularities of Complete Intersections. XIV, 153 pages. 1987.

Vol. 1294: M. Queffélec, Substitution Dynamical Systems – Spectral Analysis. XIII, 240 pages. 1987.

Vol. 1295: P. Lelong, P. Dolbeault, H. Skoda (Réd.), Séminaire d'Analyse P. Lelong – P. Dolbeault – H. Skoda. Seminar, 1985/1986. VII, 283 pages. 1987.

Vol. 1296: M.-P. Malliavin (Ed.), Séminaire d'Algèbre Paul Dubreil et Marie-Paule Malliavin. Proceedings, 1986. IV, 324 pages. 1987.

Vol. 1297: Zhu Y.-l., Guo B.-y. (Eds.), Numerical Methods for Partial Differential Equations. Proceedings. XI, 244 pages. 1987.

Vol. 1298: J. Aguadé, R. Kane (Eds.), Algebraic Topology, Barcelona 1986. Proceedings. X, 255 pages. 1987.

Vol. 1299: S. Watanabe, Yu.V. Prokhorov (Eds.), Probability Theory and Mathematical Statistics. Proceedings, 1986. VIII, 589 pages. 1988.

Vol. 1300: G.B. Seligman, Constructions of Lie Algebras and their Modules. VI, 190 pages. 1988.

Vol. 1301: N. Schappacher, Periods of Hecke Characters. XV, 160 pages. 1988.

Vol. 1302: M. Cwikel, J. Peetre, Y. Sagher, H. Wallin (Eds.), Function Spaces and Applications. Proceedings, 1986. VI, 445 pages. 1988.

Vol. 1303: L. Accardi, W. von Waldenfels (Eds.), Quantum Probability and Applications III. Proceedings, 1987. VI, 373 pages. 1988.

Vol. 1304: F.Q. Gouvêa, Arithmetic of p-adic Modular Forms. VIII, 121 pages. 1988.

Vol. 1305: D.S. Lubinsky, E.B. Saff, Strong Asymptotics for Extremal Polynomials Associated with Weights on R. VII, 153 pages. 1988.

Vol. 1306: S.S. Chern (Ed.), Partial Differential Equations. Proceedings, 1986. VI, 294 pages. 1988.

Vol. 1307: T. Murai, A Real Variable Method for the Cauchy Transform, and Analytic Capacity. VIII, 133 pages. 1988.

Vol. 1308: P. Imkeller, Two-Parameter Martingales and Their Quadratic Variation. IV, 177 pages. 1988.

Vol. 1309: B. Fiedler, Global Bifurcation of Periodic Solutions with Symmetry. VIII, 144 pages. 1988.

Vol. 1310: O.A. Laudal, G. Pfister: Local Moduli and Singularities. V, 117 pages. 1988.

Vol. 1311: A. Holme, R. Speiser (Eds.), Algebraic Geometry, Sundance 1986. Proceedings. VI, 320 pages. 1988.

Vol. 1312: N.A. Shirokov, Analytic Functions Smooth up to the Boundary. III, 213 pages. 1988.

Vol. 1313: F. Colonius. Optimal Periodic Control. VI, 177 pages. 1988.

Vol. 1314: A. Futaki. Kähler-Einstein Metrics and Integral Invariants. IV, 140 pages. 1988.

Vol. 1315: R.A. McCoy, I. Ntantu, Topological Properties of Spaces of Continuous, Functions. IV, 124 pages. 1988.

Vol. 1316: H. Korezlioglu, A.S. Ustunel (Eds.), Stochastic Analysis and Related Topics. Proceedings, 1986. V, 371 pages. 1988.

Vol. 1317: J. Lindenstrauss, V.D. Milman (Eds.), Geometric Aspects of Functional Analysis. Seminar, 1986–87. VII, 289 pages. 1988.

Vol. 1318: Y. Felix (Ed.), Algebraic Topology – Rational Homotopy. Proceedings, 1986. VIII, 245 pages. 1988.

Vol. 1319: M. Vuorinen, Conformal Geometry and Quasiregular Mappings. XIX, 209 pages. 1988.

Vol. 1320: H. Jürgensen, G. Lallement, H.J. Weinert (Eds.), Semigroups, Theory and Applications. Proceedings, 1986. X, 416 pages. 1988.

Vol. 1321: J. Azéma, P.A. Meyer, M. Yor (Eds.), Séminaire de Probabilités XXII. Proceedings. IV, 600 pages. 1988.

Vol. 1322: M. Métivier, S. Watanabe (Eds.), Stochastic Analysis. Proceedings, 1987. VII, 197 pages. 1988.

Vol. 1323: D.R. Anderson, H.J. Munkholm, Boundedly Controlled Topology. XII, 309 pages. 1988.

Vol. 1324: F. Cardoso, D.G. de Figueiredo, R. Iório, O. Lopes (Eds.), Partial Differential Equations. Proceedings, 1986. VIII, 433 pages. 1988.

Vol. 1325: A. Truman, I.M. Davies (Eds.), Stochastic Mechanics and Stochastic Processes. Proceedings, 1986. V, 220 pages. 1988.

Vol. 1326: P.S. Landweber (Ed.), Elliptic Curves and Modular Forms in Algebraic Topology. Proceedings, 1986. V, 224 pages. 1988.

Vol. 1327: W. Bruns, U. Vetter, Determinantal Rings. VII, 236 pages. 1988.

Vol. 1328: J.L. Bueso, P. Jara, B. Torrecillas (Eds.), Ring Theory. Proceedings, 1986. IX, 331 pages. 1988.

Vol. 1329: M. Alfaro, J.S. Dehesa, F.J. Marcellan, J.L. Rubio de Francia, J. Vinuesa (Eds.): Orthogonal Polynomials and their Applications. Proceedings, 1986. XV, 334 pages. 1988.

Vol. 1330: A. Ambrosetti, F. Gori, R. Lucchetti (Eds.), Mathematical Economics. Montecatini Terme 1986. Seminar. VII, 137 pages. 1988.

Vol. 1331: R. Bamón, R. Labarca, J. Palis Jr. (Eds.), Dynamical Systems, Valparaiso 1986. Proceedings. VI, 250 pages. 1988.

Vol. 1332: E. Odell, H. Rosenthal (Eds.), Functional Analysis. Proceedings. 1986–87. V, 202 pages. 1988.

Vol. 1333: A.S. Kechris, D.A. Martin, J.R. Steel (Eds.), Cabal Seminar 81–85. Proceedings, 1981–85. V, 224 pages. 1988.

Vol. 1334: Yu.G. Borisovich, Yu.E. Gliklikh (Eds.), Global Analysis – Studies and Applications III. V, 331 pages. 1988.

Vol. 1335: F. Guillén, V. Navarro Aznar, P. Pascual-Gainza, F. Puerta, Hyperrésolutions cubiques et descente cohomologique. XII, 192 pages. 1988.

Vol. 1336: B. Helffer, Semi-Classical Analysis for the Schrödinger Operator and Applications. V, 107 pages. 1988.

Vol. 1337: E. Sernesi (Ed.), Theory of Moduli. Seminar, 1985. VIII, 232 pages. 1988.

Vol. 1338: A.B. Mingarelli, S.G. Halvorsen. Non-Oscillation Domains of Differential Equations with Two Parameters. XI, 109 pages. 1988.

Vol. 1339: T. Sunada (Ed.), Geometry and Analysis of Manifolds. Proceedings, 1987. IX, 277 pages. 1988.

Vol. 1340: S. Hildebrandt, D.S. Kinderlehrer, M. Miranda (Eds.), Calculus of Variations and Partial Differential Equations. Proceedings, 1986. IX, 301 pages. 1988.

Vol. 1341: M. Dauge, Elliptic Boundary Value Problems on Corner Domains. VIII. 259 pages. 1988.

Vol. 1342: J.C. Alexander (Ed.), Dynamical Systems. Proceedings, 1986–87. VIII, 726 pages. 1988.

Vol. 1343: H. Ulrich, Fixed Point Theory of Parametrized Equivariant Maps. VII, 147 pages. 1988.

Vol. 1344: J. Král, J. Lukes, J. Netuka, J. Vesely´ (Eds.), Potential Theory – Surveys and Problems. Proceedings, 1987. VIII, 271 pages. 1988.

Vol. 1345: X. Gomez-Mont, J. Seade, A. Verjovski (Eds.), Holomorphic Dynamics. Proceedings, 1986. VII. 321 pages. 1988.

Vol. 1346: O.Ya. Viro (Ed.), Topology and Geometry – Rohlin Seminar. XI, 581 pages. 1988.

Vol. 1347: C. Preston, Iterates of Piecewise Monotone Mappings on an Interval. V, 166 pages. 1988.

Vol. 1348: F. Borceux (Ed.), Categorical Algebra and its Applications. Proceedings, 1987. VIII, 375 pages. 1988.

Vol. 1349: E. Novak, Deterministic and Stochastic Error Bounds in Numerical Analysis. V, 113 pages. 1988.

Vol. 1350: U. Koschorke (Ed.), Differential Topology Proceedings, 1987, VI, 269 pages. 1988.

Vol. 1351: I. Laine, S. Rickman, T. Sorvali (Eds.), Complex Analysis, Joensuu 1987. Proceedings. XV, 378 pages. 1988.

Vol. 1352: L.L. Avramov, K.B. Tchakerian (Eds.), Algebra – Some Current Trends. Proceedings. 1986. IX, 240 Seiten. 1988.

Vol. 1353: R.S. Palais, Ch.-l. Teng, Critical Point Theory and Submanifold Geometry. X, 272 pages. 1988.

Vol. 1354: A. Gómez, F. Guerra, M.A. Jiménez, G. López (Eds.), Approximation and Optimization. Proceedings, 1987. VI, 280 pages. 1988.

Vol. 1355: J. Bokowski, B. Sturmfels, Computational Synthetic Geometry. V, 168 pages. 1989.

Vol. 1356: H. Volkmer, Multiparameter Eigenvalue Problems and Expansion Theorems. VI, 157 pages. 1988.

Vol. 1357: S. Hildebrandt, R. Leis (Eds.), Partial Differential Equations and Calculus of Variations. VI, 423 pages. 1988.

Vol. 1358: D. Mumford, The Red Book of Varieties and Schemes. V, 309 pages. 1988.

Vol. 1359: P. Eymard, J.-P. Pier (Eds.) Harmonic Analysis. Proceedings, 1987. VIII, 287 pages. 1988.

Vol. 1360: G. Anderson, C. Greengard (Eds.), Vortex Methods. Proceedings, 1987. V, 141 pages. 1988.

Vol. 1361: T. tom Dieck (Ed.), Algebraic Topology and Transformation Groups. Proceedings. 1987. VI, 298 pages. 1988.

Vol. 1362: P. Diaconis, D. Elworthy, H. Föllmer, E. Nelson, G.C. Papanicolaou, S.R.S. Varadhan. École d´ Été de Probabilités de Saint-Flour XV–XVII. 1985–87 Editor: P.L. Hennequin. V, 459 pages. 1988.

Vol. 1363: P.G. Casazza, T.J. Shura, Tsirelson's Space. VIII, 204 pages. 1988.

Vol. 1364: R.R. Phelps, Convex Functions, Monotone Operators and Differentiability. IX, 115 pages. 1989.

Vol. 1365: M. Giaquinta (Ed.), Topics in Calculus of Variations. Seminar, 1987. X, 196 pages. 1989.

Vol. 1366: N. Levitt, Grassmannians and Gauss Maps in PL-Topology. V, 203 pages. 1989.

Vol. 1367: M. Knebusch, Weakly Semialgebraic Spaces. XX, 376 pages. 1989.

Vol. 1368: R. Hübl, Traces of Differential Forms and Hochschild Homology. III, 111 pages. 1989.

Vol. 1369: B. Jiang, Ch.-K. Peng, Z. Hou (Eds.), Differential Geometry and Topology. Proceedings, 1986–87. VI, 366 pages. 1989.

Vol. 1370: G. Carlsson, R.L. Cohen, H.R. Miller, D.C. Ravenel (Eds.), Algebraic Topology. Proceedings, 1986. IX, 456 pages. 1989.

Vol. 1371: S. Glaz, Commutative Coherent Rings. XI, 347 pages. 1989.

Vol. 1372: J. Azéma, P.A. Meyer, M. Yor (Eds.), Séminaire de Probabilités XXIII. Proceedings. IV, 583 pages. 1989.

Vol. 1373: G. Benkart, J.M. Osborn (Eds.), Lie Algebras. Madison 1987. Proceedings. V, 145 pages. 1989.

Vol. 1374: R.C. Kirby, The Topology of 4-Manifolds. VI, 108 pages. 1989.

Vol. 1375: K. Kawakubo (Ed.), Transformation Groups. Proceedings, 1987. VIII, 394 pages, 1989.

Vol. 1376: J. Lindenstrauss, V.D. Milman (Eds.), Geometric Aspects of Functional Analysis. Seminar (GAFA) 1987–88. VII, 288 pages. 1989.

Vol. 1377: J.F. Pierce, Singularity Theory, Rod Theory, and Symmetry-Breaking Loads. IV, 177 pages. 1989.

Vol. 1378: R.S. Rumely, Capacity Theory on Algebraic Curves. III, 437 pages. 1989.

Vol. 1379: H. Heyer (Ed.), Probability Measures on Groups IX. Proceedings, 1988. VIII, 437 pages. 1989.

Vol. 1380: H.P. Schlickewei, E. Wirsing (Eds.), Number Theory, Ulm 1987. Proceedings. V, 266 pages. 1989.

Vol. 1381: J.-O. Strömberg, A. Torchinsky, Weighted Hardy Spaces. V, 193 pages. 1989.

Vol. 1382: H. Reiter, Metaplectic Groups and Segal Algebras. XI, 128 pages. 1989.

Vol. 1383: D.V. Chudnovsky, G.V. Chudnovsky, H. Cohn, M.B. Nathanson (Eds.), Number Theory, New York 1985–88. Seminar. V, 256 pages. 1989.

Vol. 1384: J. Garcia-Cuerva (Ed.), Harmonic Analysis and Partial Differential Equations. Proceedings, 1987. VII, 213 pages. 1989.

Vol. 1385: A.M. Anile, Y. Choquet-Bruhat (Eds.), Relativistic Fluid Dynamics. Seminar, 1987. V, 308 pages. 1989.

Vol. 1386: A. Bellen, C.W. Gear, E. Russo (Eds.), Numerical Methods for Ordinary Differential Equations. Proceedings, 1987. VII, 136 pages. 1989.

Vol. 1387: M. Petkovi´c, Iterative Methods for Simultaneous Inclusion of Polynomial Zeros. X, 263 pages. 1989.

Vol. 1388: J. Shinoda, T.A. Slaman, T. Tugué (Eds.), Mathematical Logic and Applications. Proceedings, 1987. V, 223 pages. 1989.

Vol. 1000: Second Edition. H. Hopf, Differential Geometry in the Large. VII, 184 pages. 1989.

Vol. 1389: E. Ballico, C. Ciliberto (Eds.), Algebraic Curves and Projective Geometry. Proceedings, 1988. V, 288 pages. 1989.

Vol. 1390: G. Da Prato, L. Tubaro (Eds.), Stochastic Partial Differential Equations and Applications II. Proceedings, 1988. VI, 258 pages. 1989.

Vol. 1391: S. Cambanis, A. Weron (Eds.), Probability Theory on Vector Spaces IV. Proceedings, 1987. VIII, 424 pages. 1989.

Vol. 1392: R. Silhol, Real Algebraic Surfaces. X, 215 pages. 1989.

Vol. 1393: N. Bouleau, D. Feyel, F. Hirsch, G. Mokobodzki (Eds.), Séminaire de Théorie du Potentiel Paris, No. 9. Proceedings. VI, 265 pages. 1989.

Vol. 1394: T.L. Gill, W.W. Zachary (Eds.), Nonlinear Semigroups, Partial Differential Equations and Attractors. Proceedings, 1987. IX, 233 pages. 1989.

Vol. 1395: K. Alladi (Ed.), Number Theory, Madras 1987. Proceedings. VII, 234 pages. 1989.

Vol. 1396: L. Accardi, W. von Waldenfels (Eds.), Quantum Probability and Applications IV. Proceedings, 1987. VI, 355 pages. 1989.

Vol. 1397: P.R. Turner (Ed.), Numerical Analysis and Parallel Processing. Seminar, 1987. VI, 264 pages. 1989.

Vol. 1398: A.C. Kim, B.H. Neumann (Eds.), Groups – Korea 1988. Proceedings. V, 189 pages. 1989.

Vol. 1399: W.-P. Barth, H. Lange (Eds.), Arithmetic of Complex Manifolds. Proceedings, 1988. V, 171 pages. 1989.

Vol. 1400: U. Jannsen. Mixed Motives and Algebraic K-Theory. XIII, 246 pages. 1990.

Vol. 1401: J. Steprans, S. Watson (Eds.), Set Theory and its Applications. Proceedings, 1987. V, 227 pages. 1989.

Vol. 1402: C. Carasso, P. Charrier, B. Hanouzet, J.-L. Joly (Eds.), Nonlinear Hyperbolic Problems. Proceedings, 1988. V, 249 pages. 1989.

Vol. 1403: B. Simeone (Ed.), Combinatorial Optimization. Seminar, 1986. V, 314 pages. 1989.

Vol. 1404: M.-P. Malliavin (Ed.), Séminaire d´Algèbre Paul Dubreil et Marie-Paul Malliavin. Proceedings, 1987–1988. IV, 410 pages. 1989.

Vol. 1405: S. Dolecki (Ed.), Optimization. Proceedings, 1988. V, 223 pages. 1989. Vol. 1406: L. Jacobsen (Ed.), Analytic Theory of Continued Fractions III. Proceedings, 1988. VI, 142 pages. 1989.

Vol. 1407: W. Pohlers, Proof Theory. VI, 213 pages. 1989.

Vol. 1408: W. Lück, Transformation Groups and Algebraic K-Theory. XII, 443 pages. 1989.

Vol. 1409: E. Hairer, Ch. Lubich, M. Roche. The Numerical Solution of Differential-Algebraic Systems by Runge-Kutta Methods. VII, 139 pages. 1989.

Vol. 1410: F.J. Carreras, O. Gil-Medrano, A.M. Naveira (Eds.), Differential Geometry. Proceedings, 1988. V, 308 pages. 1989.

Vol. 1411: B. Jiang (Ed.), Topological Fixed Point Theory and Applications. Proceedings. 1988. VI, 203 pages. 1989.

Vol. 1412: V.V. Kalashnikov, V.M. Zolotarev (Eds.), Stability Problems for Stochastic Models. Proceedings, 1987. X, 380 pages. 1989.

Vol. 1413: S. Wright, Uniqueness of the Injective III_1 Factor. III, 108 pages. 1989.

Vol. 1414: E. Ramirez de Arellano (Ed.), Algebraic Geometry and Complex Analysis. Proceedings, 1987. VI, 180 pages. 1989.

Vol. 1415: M. Langevin, M. Waldschmidt (Eds.), Cinquante Ans de Polynômes. Fifty Years of Polynomials. Proceedings, 1988. IX, 235 pages.1990.

Vol. 1416: C. Albert (Ed.), Géométrie Symplectique et Mécanique. Proceedings, 1988. V, 289 pages. 1990.

Vol. 1417: A.J. Sommese, A. Biancofiore, E.L. Livorni (Eds.), Algebraic Geometry. Proceedings, 1988. V, 320 pages. 1990.

Vol. 1418: M. Mimura (Ed.), Homotopy Theory and Related Topics. Proceedings, 1988. V, 241 pages. 1990.

Vol. 1419: P.S. Bullen, P.Y. Lee, J.L. Mawhin, P. Muldowney, W.F. Pfeffer (Eds.), New Integrals. Proceedings, 1988. V, 202 pages. 1990.

Vol. 1420: M. Galbiati, A. Tognoli (Eds.), Real Analytic Geometry. Proceedings, 1988. IV, 366 pages. 1990.

Vol. 1421: H.A. Biagioni, A Nonlinear Theory of Generalized Functions, XII, 214 pages. 1990.

Vol. 1422: V. Villani (Ed.), Complex Geometry and Analysis. Proceedings, 1988. V, 109 pages. 1990.

Vol. 1423: S.O. Kochman, Stable Homotopy Groups of Spheres: A Computer-Assisted Approach. VIII, 330 pages. 1990.

Vol. 1424: F.E. Burstall, J.H. Rawnsley, Twistor Theory for Riemannian Symmetric Spaces. III, 112 pages. 1990.

Vol. 1425: R.A. Piccinini (Ed.), Groups of Self-Equivalences and Related Topics. Proceedings, 1988. V, 214 pages. 1990.

Vol. 1426: J. Azéma, P.A. Meyer, M. Yor (Eds.), Séminaire de Probabilités XXIV, 1988/89. V, 490 pages. 1990.

Vol. 1427: A. Ancona, D. Geman, N. Ikeda, École d'Eté de Probabilités de Saint Flour

XVIII, 1988. Ed.: P.L. Hennequin. VII, 330 pages. 1990.

Vol. 1428: K. Erdmann, Blocks of Tame Representation Type and Related Algebras. XV. 312 pages. 1990.

Vol. 1429: S. Homer, A. Nerode, R.A. Platek, G.E. Sacks, A. Scedrov, Logic and Computer Science. Seminar, 1988. Editor: P. Odifreddi. V, 162 pages. 1990.

Vol. 1430: W. Bruns, A. Simis (Eds.), Commutative Algebra. Proceedings. 1988. V, 160 pages. 1990.

Vol. 1431: J.G. Heywood, K. Masuda, R. Rautmann, V.A. Solonnikov (Eds.), The Navier-Stokes Equations – Theory and Numerical Methods. Proceedings, 1988. VII, 238 pages. 1990.

Vol. 1432: K. Ambos-Spies, G.H. Müller, G.E. Sacks (Eds.), Recursion Theory Week. Proceedings, 1989. VI, 393 pages. 1990.

Vol. 1433: S. Lang, W. Cherry, Topics in Nevanlinna Theory. II, 174 pages.1990.

Vol. 1434: K. Nagasaka, E. Fouvry (Eds.), Analytic Number Theory. Proceedings, 1988. VI, 218 pages. 1990.

Vol. 1435: St. Ruscheweyh, E.B. Saff, L.C. Salinas, R.S. Varga (Eds.), Computational Methods and Function Theory. Proceedings, 1989. VI, 211 pages. 1990.

Vol. 1436: S. Xambó-Descamps (Ed.), Enumerative Geometry. Proceedings, 1987. V, 303 pages. 1990.

Vol. 1437: H. Inassaridze (Ed.), K-theory and Homological Algebra. Seminar, 1987–88. V, 313 pages. 1990.

Vol. 1438: P.G. Lemarié (Ed.) Les Ondelettes en 1989. Seminar. IV, 212 pages. 1990.

Vol. 1439: E. Bujalance, J.J. Etayo, J.M. Gamboa, G. Gromadzki. Automorphism Groups of Compact Bordered Klein Surfaces: A Combinatorial Approach. XIII, 201 pages. 1990.

Vol. 1440: P. Latiolais (Ed.), Topology and Combinatorial Groups Theory. Seminar, 1985–1988. VI, 207 pages. 1990.

Vol. 1441: M. Coornaert, T. Delzant, A. Papadopoulos. Géométrie et théorie des groupes. X, 165 pages. 1990.

Vol. 1442: L. Accardi, M. von Waldenfels (Eds.), Quantum Probability and Applications V. Proceedings, 1988. VI, 413 pages. 1990.

Vol. 1443: K.H. Dovermann, R. Schultz, Equivariant Surgery Theories and Their Periodicity Properties. VI, 227 pages. 1990.

Vol. 1444: H. Korezlioglu, A.S. Ustunel (Eds.), Stochastic Analysis and Related Topics VI. Proceedings, 1988. V, 268 pages. 1990.

Vol. 1445: F. Schulz, Regularity Theory for Quasilinear Elliptic Systems and – Monge Ampère Equations in Two Dimensions. XV, 123 pages. 1990.

Vol. 1446: Methods of Nonconvex Analysis. Seminar, 1989. Editor: A. Cellina. V, 206 pages. 1990.

Vol. 1447: J.-G. Labesse, J. Schwermer (Eds), Cohomology of Arithmetic Groups and Automorphic Forms. Proceedings, 1989. V, 358 pages. 1990.

Vol. 1448: S.K. Jain, S.R. López-Permouth (Eds.), Non-Commutative Ring Theory. Proceedings, 1989. V, 166 pages. 1990.

Vol. 1449: W. Odyniec, G. Lewicki, Minimal Projections in Banach Spaces. VIII, 168 pages. 1990.

Vol. 1450: H. Fujita, T. Ikebe, S.T. Kuroda (Eds.), Functional-Analytic Methods for Partial Differential Equations. Proceedings, 1989. VII, 252 pages. 1990.

Vol. 1451: L. Alvarez-Gaumé, E. Arbarello, C. De Concini, N.J. Hitchin, Global Geometry and Mathematical Physics. Montecatini Terme 1988. Seminar. Editors: M. Francaviglia, F. Gherardelli. IX, 197 pages. 1990.

Vol. 1452: E. Hlawka, R.F. Tichy (Eds.), Number-Theoretic Analysis. Seminar, 1988–89. V, 220 pages. 1990.

Vol. 1453: Yu.G. Borisovich, Yu.E. Gliklikh (Eds.), Global Analysis – Studies and Applications IV. V, 320 pages. 1990.

Vol. 1454: F. Baldassari, S. Bosch, B. Dwork (Eds.), p-adic Analysis. Proceedings, 1989. V, 382 pages. 1990.

Vol. 1455: J.-P. Françoise, R. Roussarie (Eds.), Bifurcations of Planar Vector Fields. Proceedings, 1989. VI, 396 pages. 1990.

Vol. 1456: L.G. Kovács (Ed.), Groups – Canberra 1989. Proceedings. XII, 198 pages. 1990.

Vol. 1457: O. Axelsson, L.Yu. Kolotilina (Eds.), Preconditioned Conjugate Gradient Methods. Proceedings, 1989. V, 196 pages. 1990.

Vol. 1458: R. Schaaf, Global Solution Branches of Two Point Boundary Value Problems. XIX, 141 pages. 1990.

Vol. 1459: D. Tiba, Optimal Control of Nonsmooth Distributed Parameter Systems. VII, 159 pages. 1990.

Vol. 1460: G. Toscani, V. Boffi, S. Rionero (Eds.), Mathematical Aspects of Fluid Plasma Dynamics. Proceedings, 1988. V, 221 pages. 1991.

Vol. 1461: R. Gorenflo, S. Vessella, Abel Integral Equations. VII, 215 pages. 1991.

Vol. 1462: D. Mond, J. Montaldi (Eds.), Singularity Theory and its Applications. Warwick 1989, Part I. VIII, 405 pages. 1991.

Vol. 1463: R. Roberts, I. Stewart (Eds.), Singularity Theory and its Applications. Warwick 1989, Part II. VIII, 322 pages. 1991.

Vol. 1464: D. L. Burkholder, E. Pardoux, A. Sznitman, Ecole d'Eté de Probabilités de Saint- Flour XIX-1989. Editor: P. L. Hennequin. VI, 256 pages. 1991.

Vol. 1465: G. David, Wavelets and Singular Integrals on Curves and Surfaces. X, 107 pages. 1991.

Vol. 1466: W. Banaszczyk, Additive Subgroups of Topological Vector Spaces. VII, 178 pages. 1991.